International Vocational Education Bilingual Textbook Series
国际化职业教育双语系列教材

Steelmaking Technology
炼钢生产技术

Li Xiujuan   Jia Yanlu
李秀娟   贾燕璐   编

Beijing
Metallurgical Industry Press
2020

内 容 提 要

本书共分4个项目，主要内容包括基础知识培训、生产准备、电炉炼钢生产以及电炉炉衬的砌筑和维护。

本书既可作为职业院校冶金相关专业的国际化教学用书，也可作为冶金企业员工的培训教材和有关专业人员的参考书。

图书在版编目(CIP)数据

炼钢生产技术＝Steelmaking Technology：汉、英／李秀娟，贾燕璐编 . —北京：冶金工业出版社，2020.8
国际化职业教育双语系列教材
ISBN 978-7-5024-8534-4

Ⅰ.①炼… Ⅱ.①李… ②贾… Ⅲ.①转炉炼钢—高等职业教育—双语教学—教材—汉、英 Ⅳ.①TF71

中国版本图书馆 CIP 数据核字(2020)第 152040 号

出 版 人　陈玉千
地　　址　北京市东城区嵩祝院北巷39号　邮编　100009　电话　(010)64027926
网　　址　www.cnmip.com.cn　电子信箱　yjcbs@cnmip.com.cn
责任编辑　俞跃春　杜婷婷　美术编辑　郑小利　版式设计　孙跃红　禹蕊
责任校对　李　娜　责任印制　李玉山
ISBN 978-7-5024-8534-4

冶金工业出版社出版发行；各地新华书店经销；三河市双峰印刷装订有限公司印刷
2020年8月第1版，2020年8月第1次印刷
787mm×1092mm　1/16；13印张；291千字；186页
49.00元

冶金工业出版社　投稿电话　(010)64027932　投稿信箱　tougao@cnmip.com.cn
冶金工业出版社营销中心　电话　(010)64044283　传真　(010)64027893
冶金工业出版社天猫旗舰店　yjgycbs.tmall.com

(本书如有印装质量问题，本社营销中心负责退换)

Editorial Board of International Vocational Education Bilingual Textbook Series

Director Kong Weijun (Party Secretary and Dean of Tianjin Polytechnic College)

Deputy Director Zhang Zhigang (Chairman of Tiantang Group, Sino-Uganda Mbale Industrial Park)

Committee Members Li Guiyun, Li Wenchao, Zhao Zhichao, Liu Jie, Zhang Xiufang, Tan Qibing, Liang Guoyong, Zhang Tao, Li Meihong, Lin Lei, Ge Huijie, Wang Zhixue, Wang Xiaoxia, Li Rui, Yu Wansong, Wang Lei, Gong Na, Li Xiujuan, Zhang Zhichao, Yue Gang, Xuan Jie, Liang Luan, Chen Hong, Jia Yanlu, Chen Baoling

国际化职业教育双语系列教材编委会

主　任　孔维军（天津工业职业学院党委书记、院长）

副主任　张志刚（中乌姆巴莱工业园天唐集团董事长）

委　员　李桂云　李文潮　赵志超　刘　洁　张秀芳
　　　　　谭起兵　梁国勇　张　涛　李梅红　林　磊
　　　　　葛慧杰　王治学　王晓霞　李　蕊　于万松
　　　　　王　磊　宫　娜　李秀娟　张志超　岳　刚
　　　　　玄　洁　梁　娈　陈　红　贾燕璐　陈宝玲

Foreword

With the proposal of the 'Belt and Road Initiative', the Ministry of Education of China issued *Promoting Education Action for Building the Belt and Road Initiative* in 2016, proposing cooperation in education, including 'cooperation in human resources training'. At the Forum on China-Africa Cooperation (FOCAC) in 2018, President Xi proposed to focus on the implementation of the 'Eight Actions', which put forward the plan to establish 10 Luban Workshops to provide skills training to African youth. Draw lessons from foreign advanced experience of vocational education mode, China's vocational education has continuously explored and formed the new mode of vocational education with Chinese characteristics. Tianjin, as a demonstration zone for reform and innovation of modern vocational education in China, has started the construction of 'Luban Workshop' along the 'Belt and Road Initiative', to export high-quality vocational education achievements.

The compilation of these series of textbooks is in response to the times and it's also the beginning of Tianjin Polytechnic College to explore the internationalization of higher vocational education. It's a new model of vocational education internationalization by Tianjin, response to the 'Belt and Road Initiative' and the 'Going Out' of Chinese enterprises. Tianjin Polytechnic College and Uganda Technical College-Elgon reached a cooperation intention to establish the Luban Workshop to carry out vocational education cooperation on mechatronics technology and ferrous metallurgy technology major in 2019. The establishment of Luban Workshop is conducive to strengthen the cooperation between China and Uganda in vocational education, promote the export of high-quality higher vocational education resources, and serve Chinese enterprises in Uganda and Ugandan local enterprises. Exploring and standardizing the overseas operation of Chinese colleges, the expansion of international influences of China's higher vocational education is also one of the purposes.

The construction of 'Luban Workshop' in Uganda is mainly based on the EPIP (Engineering, Practice, Innovation, Project) project, and is committed to cultivating high-quality talents with innovative spirit, creative ability and entrepreneurial spirit. To meet the learning needs of local teachers and students accurately, the compilation of these international vocational skills bilingual textbooks is based on the talent demand of Uganda and the specialty and characteristics of Tianjin Polytechnic.

These textbooks are supporting teaching material, referring to Chinese national professional standards and developing international professional teaching standards. The internationalization of the curriculums takes into account the technical skills and cognitive characteristics of local students, to promote students' communication and learning ability. At the same time, these textbooks focus on the enhancement of vocational ability, rely on professional standards, and integrate the teaching concept of equal emphasis on skills and quality. These textbooks also adopted project-based, modular, task-driven teaching model and followed the requirements of enterprise posts for employees.

In the process of writing the series of textbooks, Wang Xiaoxia, Li Rui, Wang Zhixue, Ge Huijie, Yu Wansong, Wang Lei, Li Xiujuan, Gong Na, Zhang Zhichao, Jia Yanlu, Chen Baoling and other chief teachers, professional teams, English teaching and research office have made great efforts, receiving strong support from leaders of Tianjin Polytechnic College. During the compilation, the series of textbooks referred to a large number of research findings of scholars in the field, and we would like to thank them for their contributions.

Finally, we sincerely hope that the series of textbooks can contribute to the internationalization of China's higher vocational education, especially to the development of higher vocational education in Africa.

Principal of Tianjin Polytechnic College Kong Weijun
May, 2020

序

随着"一带一路"倡议的提出，2016年中华人民共和国教育部发布了《推进共建"一带一路"教育行动》，提出了包括"开展人才培养培训合作"在内的教育合作。2018年习近平主席在中非合作论坛上提出，要重点实施"八大行动"，明确要求在非洲设立10个鲁班工坊，向非洲青年提供技能培训。中国职业教育在吸收和借鉴发达国家先进职教发展模式的基础上，不断探索和形成了中国特色职业教育办学模式。天津市作为中国现代职业教育改革创新示范区，开启了"鲁班工坊"建设工作，在"一带一路"沿线国家搭建"鲁班工坊"平台，致力于把优秀职业教育成果输出国门与世界分享。

本系列教材的编写，契合时代大背景，是天津工业职业学院探索高职教育国际化的开端。"鲁班工坊"是由天津率先探索和构建的一种职业教育国际化发展新模式，是响应国家"一带一路"倡议和中国企业"走出去"，创建职业教育国际合作交流的新窗口。2019年天津工业职业学院与乌干达埃尔贡技术学院达成合作意向，共同建立"鲁班工坊"，就机电一体化技术专业、黑色冶金技术专业开展职业教育合作。此举旨在加强中乌职业教育交流与合作，推动中国优质高等职业教育资源"走出去"，服务在乌中资企业和乌干达当地企业，探索和规范我国职业院校"鲁班工坊"建设和境外办学，扩大中国高等职业教育的国际影响力。

中乌"鲁班工坊"的建设主要以工程实践创新项目（EPIP：Engineering，Practice，Innovation，Project）为载体，致力于培养具有创新精神、创造能力和创业精神的"三创"复合型高素质技能人才。国际化职业教育双语系列教材的编写，立足于乌干达人才需求和天津工业职业学院专业特色，是为了更好满足当地师生学习需求。

本系列教材采用中英双语相结合的方式，主要参照中国专业标准，开发国际化专业教学标准，课程内容国际化是在专业课程设置上，结合本地学生的技术能力水平与认知特点，合理设置双语教学环节，加强学生的学习与交流能力。同时，教材以提升职业能力为核心，以职业标准为依托，体现技能与质量并重的教学理念，主要采用项目化、模块化、任务驱动的教学模式，并结合企业岗位对员工的要求来撰写。

本系列教材在撰写过程中，王晓霞、李蕊、王治学、葛慧杰、于万松、王磊、李秀娟、宫娜、张志超、贾燕璐、陈宝玲等主编老师、专业团队、英语教研室付出了辛勤劳动，并得到了学院各级领导的大力支持，同时本系列教材借鉴和参考了业界有关学者的研究成果，在此一并致谢！

最后，衷心希望本系列教材能为我国高等职业教育国际化，尤其是高等职业教育走进非洲、支援非洲高等职业教育发展尽绵薄之力。

天津工业职业学院书记、院长　孔维军
2020 年 5 月

Preface

Tianjin Polytechnic College and Uganda Technical College-Elgon reached a cooperation intention to establish the Luban Workshop to carry out vocational education cooperation on mechatronics technology and ferrous metallurgy technology major in 2019. In order to strengthen the cooperation between China and Uganda in vocational education, the two colleges plan to compile a series of international vocational skills bilingual textbooks.

This book is one of the international vocational skills bilingual textbooks. According to the characteristics of the course design, such as wide range of subjects and strong practicality, this book introduces the production technology of EAF steelmaking in a simple way, avoids the tedious theoretical derivation, enhances the practical application, and comprehensively introduces the production process and technology of steelmaking. This book introduces the new achievements and progress in the field of EAF steelmaking at home and abroad as far as possible, and fully embodies the purpose of higher vocational education to cultivate application-oriented talents. This book is compiled with reference to the national vocational standards of electric furnace steelmaking industry, in accordance with the requirements of enterprise posts for employees, based on the vocational standards, with the improvement of professional ability as the core, with the focus on the operation of the work process, and using the task driven teaching mode.

This book takes the production process and equipment of electric furnace steelmaking as the carrier, mainly introduces the theoretical knowledge and practical skills of the production equipment of electric furnace steelmaking, operation process of electric furnace steelmaking, lining masonry and maintenance of electric furnace, etc. This book is the result of school enterprise cooperation. The editor, together with the technical experts in the production line, tracks the technical development trend on the basis of the industry experts and graduates' job research, and compiles this book according to the cultivation of application-oriented talents in the electric furnace steel-making enterprises. This book was jointly completed by Li Xiujuan and Jia Yanlu of Tianjin Institute of technology. Many of the author's documents are cited in the book, and I would like to express my thanks.

Due to the limited level of editors, it is inevitable that there are some defects or omissions in the book. Please criticize and correct them.

The editor
April, 2020

前　言

2019年天津工业职业学院与乌干达埃尔贡技术学院达成合作意向，共同建立"鲁班工坊"，就机电一体化技术专业、黑色冶金技术专业开展职业教育合作，双方计划编撰国际化职业教育双语系列教材。

本书是国际化职业教育双语系列教材之一。本书根据系列教材要求和课程设计的学科面广、实践性强等特点，深入浅出地介绍了电炉炼钢生产技术，避免了繁琐的理论推导，增强了实际应用，对炼钢的生产工艺与技术进行了全面的介绍。本书尽可能地介绍了国内外电炉炼钢生产领域的新成果、新进展，充分体现高等职业教育培养应用型人才的宗旨。本书参照电炉炼钢工国家职业标准，遵循企业岗位对就业人员的要求，以职业标准为依托，以提升职业能力为核心，以工作过程操作为重点，采用任务驱动的教学模式来进行编写。

本书以电炉炼钢生产工艺及设备为载体，主要介绍了电炉炼钢生产设备、电炉炼钢生产操作流程、电炉炉衬砌筑及维护等方面的理论知识和实践技能。本书由天津工业职业学院李秀娟和贾燕璐共同完成。本书在编写过程中参考了有关文献资料，在此对相关作者表示衷心的感谢。

由于编者水平所限，书中不妥之处，恳请读者批评指正。

编　者
2020年4月

Contents

Project 1   Basic Knowledge Training 1

 Task 1.1   Cognition of EAF Steelmaking Production 1
  1.1.1   Classification of Smelting Methods 2
  1.1.2   Principle and Task of EAF Steelmaking 4
 Task 1.2   Cognition of Smelting Principle of Electric Furnace 5
  1.2.1   Decarbonization 5
  1.2.2   Dephosphorization 6
  1.2.3   Desulfurization 6
 Task 1.3   Cognition of Raw Materials for Electric Furnace Smelting 7
  1.3.1   Steel Scrap 7
  1.3.2   Pig Iron 8
  1.3.3   Oxidant 9
  1.3.4   Deoxidizer and Ferroalloy 9
  1.3.5   Slagging Material 10
  1.3.6   Carburizer 10
  1.3.7   Electrode 10

Project 2   Production Preparation 13

 Task 2.1   Mechanical Equipment Preparation 13
  2.1.1   Equipment Composition 13
  2.1.2   Furnace Body 13
  2.1.3   Electrode Arms and Electrode Lifting Device 20
  2.1.4   Roof Lifting System 23
  2.1.5   Furnace Tilting Device 24
  2.1.6   Top Charging System 24
  2.1.7   Auxiliary Equipment for Electric Furnace 26
 Task 2.2   Electrical Equipment Preparation 30
  2.2.1   Basic Function and Principle of Electric Equipment 30
  2.2.2   Electric Equipment 34

Project 3 Electric Furnace Steelmaking Production ... 44

Task 3.1 Operation of Charging and Batching ... 44
3.1.1 Make Up the Stove ... 44
3.1.2 Ingredients ... 45
3.1.3 Loading ... 52

Task 3.2 Melting Period Operation ... 58
3.2.1 Melting Process of Charge ... 58
3.2.2 The Main Factors Affecting the Melting of Charge ... 60
3.2.3 Physicochemical Reaction in Melting Period ... 63
3.2.4 Dephosphorization Operation in Melting Period ... 66
3.2.5 Melting Period Operation ... 67

Task 3.3 Operation in Oxidation Period ... 68
3.3.1 Oxidation Method ... 69
3.3.2 Decarbonization Operation ... 71
3.3.3 Oxidation of Iron, Silicon, Manganese, Chromium and Other Elements ... 79
3.3.4 Establishment of Smelting Temperature System and Temperature Rise of Molten Steel ... 86
3.3.5 Full Slagging and Carburization ... 90
3.3.6 Oxidation Period Operation ... 94

Task 3.4 Restore Period Operation ... 100
3.4.1 Deoxidation ... 101
3.4.2 Desulfurization in Reduction Period ... 122
3.4.3 Adjustment and Measurement of Liquid Steel Temperature ... 126
3.4.4 Adjustment of Liquid Steel Composition ... 135

Task 3.5 Tapping Operation ... 151
3.5.1 Tapping Conditions of Electric Furnace ... 151
3.5.2 Tapping Method of Electric Furnace ... 154
3.5.3 Tapping Operation of Electric Furnace ... 154
3.5.4 Experience Judgment of Tapping Temperature ... 156

Project 4 Masonry and Maintenance of EAF Lining ... 159

Task 4.1 Main Properties and Classification of Refractories ... 159
4.1.1 Main Performance Indexes of Refractories ... 159
4.1.2 Classification of Refractories ... 161

Task 4.2 Refractories for EAF ... 162
4.2.1 General Requirements for Refractories in EAF Steelmaking ... 163
4.2.2 Refractory for Roof ... 163
4.2.3 Refractory for Wall and Bottom ... 166
4.2.4 Insulation and Binder for EAF ... 168

Task **4.3**	Masonry of Furnace Lining	171
4.3.1	Masonry of Furnace Bottom	171
4.3.2	Masonry of Furnace Walls	174
4.3.3	Masonry of Roof	175
Task **4.4**	Maintenance of Furnace Lining	177
4.4.1	Causes of Lining Damage	177
4.4.2	Furnace Bottom and Slope Maintenance	178
4.4.3	Furnace Wall Maintenance	179
4.4.4	Roof Maintenance	180
4.4.5	Water-cooled Furnace Lining	181

References 185

目 录

项目1 基础知识培训 ... 1

任务1.1 电炉炼钢生产认知 ... 1
1.1.1 冶炼方法的分类 ... 2
1.1.2 电弧炉炼钢的原理及任务 ... 4

任务1.2 电炉冶炼原理认知 ... 5
1.2.1 脱碳 ... 5
1.2.2 脱磷 ... 6
1.2.3 脱硫 ... 6

任务1.3 电炉冶炼原材料认知 ... 7
1.3.1 废钢 ... 7
1.3.2 生铁 ... 8
1.3.3 氧化剂 ... 9
1.3.4 脱氧剂和铁合金 ... 9
1.3.5 造渣材料 ... 10
1.3.6 增碳剂 ... 10
1.3.7 电极 ... 10

项目2 生产准备 ... 13

任务2.1 机械设备准备 ... 13
2.1.1 设备构成 ... 13
2.1.2 炉体 ... 13
2.1.3 电极横臂及电极升降装置 ... 20
2.1.4 炉盖提升系统 ... 23
2.1.5 炉体倾动装置 ... 24
2.1.6 炉顶装料系统 ... 24
2.1.7 电炉的辅助设备 ... 26

任务2.2 电气设备准备 ... 30
2.2.1 电弧炉的电气设备的基本作用及原理 ... 30
2.2.2 电弧炉相关电气设备 ... 34

项目3　电炉炼钢生产 ... 44

任务3.1　补炉配料装料操作 ... 44
3.1.1　补炉 ... 44
3.1.2　配料 ... 45
3.1.3　装料 ... 52

任务3.2　熔化期操作 ... 58
3.2.1　炉料的熔化过程 ... 58
3.2.2　影响炉料熔化的主要因素 ... 60
3.2.3　熔化期的物化反应 ... 63
3.2.4　熔化期脱磷操作 ... 66
3.2.5　熔化期操作 ... 67

任务3.3　氧化期及操作 ... 68
3.3.1　氧化方法 ... 69
3.3.2　脱碳操作 ... 71
3.3.3　铁、硅、锰、铬等元素的氧化 ... 79
3.3.4　冶炼温度制度的制订和钢液的升温 ... 86
3.3.5　全扒渣与增碳 ... 90
3.3.6　氧化期操作 ... 94

任务3.4　还原期操作 ... 100
3.4.1　脱氧 ... 101
3.4.2　还原期的脱硫 ... 122
3.4.3　钢液温度的调整与测量 ... 126
3.4.4　钢液成分的调整 ... 135

任务3.5　出钢操作 ... 151
3.5.1　电炉的出钢条件 ... 151
3.5.2　电炉的出钢方式 ... 154
3.5.3　电炉的出钢操作 ... 154
3.5.4　出钢温度的经验判断 ... 156

项目4　电炉炉衬的砌筑和维护 ... 159

任务4.1　耐火材料的主要性能和分类 ... 159
4.1.1　耐火材料的主要性能指标 ... 159
4.1.2　耐火材料的分类 ... 161

任务4.2　电弧炉用耐火材料 ... 162
4.2.1　电炉炼钢对耐火材料的一般要求 ... 163
4.2.2　炉盖用耐火材料 ... 163
4.2.3　炉墙、炉底用耐火材料 ... 166
4.2.4　电炉用绝热材料和黏结剂 ... 168

任务4.3 炉衬的砌筑 ·· 171
 4.3.1 炉底的砌筑 ·· 171
 4.3.2 炉墙的砌筑 ·· 174
 4.3.3 炉盖的砌筑 ·· 175
任务4.4 炉衬的维护 ·· 177
 4.4.1 炉衬损坏的原因 ··· 177
 4.4.2 炉底和炉坡的维护 ·· 178
 4.4.3 炉墙的维护 ·· 179
 4.4.4 炉盖的维护 ·· 180
 4.4.5 水冷炉衬 ··· 181

参考文献 ·· 185

Project 1　Basic Knowledge Training
项目1　基础知识培训

Task 1.1　Cognition of EAF Steelmaking Production
任务1.1　电炉炼钢生产认知

Mission objectives
任务目标

(1) Understand the basic principles of electric furnace smelting.
(1) 理解电炉冶炼基本原理。
(2) Master the production task of electric furnace steelmaking.
(2) 掌握电炉炼钢生产任务。

EAF steelmaking, which mainly refers to EAF steelmaking, is the main method to produce special steel at home and abroad. At present, more than 90% of the world's EAF steel is produced by EAF, and a small amount of EAF steel is produced by induction furnace, electroslag furnace, etc. Generally speaking, the electric arc furnace refers to the alkaline electric arc furnace. EAF steelmaking has a history of one hundred years. The first EAF to be used is DC EAF. Because of the expansion of furnace capacity, it was difficult to provide high-power rectifying power, so there was a three-phase AC EAF, which has been dominant for a long time. It is mainly used for high alloy steel and special steel. Due to its slow speed and low energy consumption, after 1960s, oxygen blowing was first used to replace ore, so that scrap steel can be used to produce wire rod and bar steel with low cost.

电炉炼钢，主要是指电弧炉炼钢，是目前国内外生产特殊钢的主要方法。目前，世界上90%以上的电炉钢是电弧炉生产的，还有少量电炉钢是由感应炉、电渣炉等生产的。通常所说的电弧炉，是指碱性电弧炉。电弧炉炼钢至今有百年历史，最先使用的电弧炉为直流电弧炉，由于炉容的扩大，当时很难提供大功率整流电源，因此出现了三相交流电弧炉，并长期占据主导地位。主要用于高合金钢和特殊钢，由于速度慢、能耗低，20世纪60年代后，首先采用吹氧代替矿石，使之可以利用废钢较低成本地生产线材、棒材用钢。

Main features of EAF steelmaking:
(1) Waste iron and steel or crude steel is the main solid material, which does not need a

huge iron making and coking system;

(2) The raw materials are widely used and are suitable for short-term and intermittent production in production scheduling;

(3) By using electric energy to melt (remelt) and heat up, it is easy to obtain high temperature, convenient and accurate temperature regulation, which is conducive to smelting all kinds of steel;

(4) The atmosphere (oxidation, reduction, neutral), vacuum degree and pressure in the furnace can be controlled;

(5) The yield of the alloy is high, the composition is easy to adjust and the control range is narrow;

(6) It has wide variety adaptability and good quality, and can smelt various types of high quality steel and alloy steel with low P, S and O content;

(7) The equipment is simple, the technological process is short, the floor area is small, the cost of capital construction is low, the production is fast, and the pollution is easy to control;

(8) High power consumption, [N], [H];

(9) The productivity is lower than that of converter, and the consumption of electrode and refractory is higher than that of other smelting methods.

电弧炉炼钢的主要特征：

（1）以废钢铁或粗钢为主要固体料，不需要一套庞大的炼铁和炼焦系统；

（2）原料使用范围广，在生产调度上适于短时间、间歇式生产；

（3）利用电能使其熔化（重熔）及升温，易获得高温，调温方便而准确，有利于冶炼各类钢种；

（4）炉内气氛（氧化、还原、中性）、真空度、压力可调控；

（5）合金收得率高，成分易于调整且控制范围窄；

（6）品种适应性广、质量好，可冶炼含P、S、O低的各种类型的优质钢和合金钢；

（7）设备较简单，工艺流程短，占地面积小，基建费用低，投产快，易控制污染；

（8）消耗电能大，[N]、[H]高；

（9）生产率比转炉低，电极、耐材等消耗比其他冶炼方法高。

1.1.1　Classification of Smelting Methods
1.1.1　冶炼方法的分类

According to the charging state, there are hot charging and cold charging. Hot charging has no melting period, short smelting time and high productivity, but it needs converter or other forms of mixer; cold charging mainly uses solid steel or sponge iron, etc. According to the slag forming times in smelting process, there are single slag method and double slag method. According to the use of oxygen and no oxygen in the smelting process, there are oxidation method and non oxidation method. The oxidation method mostly uses double slag smelting, but there are also single slag smelting, such as the rapid smelting of electric furnace steel, while the non oxidation method uses single slag smelting. In addition, there is the return oxygen blowing method. According to the different ways of oxygen supply in the oxidation period, there are ore oxidation, oxygen oxidation, ore oxygen comprehensive oxidation and argon oxygen mixed blowing.

根据炉料的入炉状态分,有热装和冷装两种。热装没有熔化期,冶炼时间短,生产率高,但需转炉或其他形式的混铁炉配合;冷装主要使用固体钢铁料或海绵铁等。根据冶炼过程中的造渣次数分,有单渣法和双渣法。根据冶炼过程中用氧与不用氧来分,有氧化法和不氧化法。氧化法多采用双渣冶炼,但也有采用单渣冶炼的,如电炉钢的快速冶炼,而不氧化法均采用单渣冶炼。此外,还有返回吹氧法。根据氧化期供氧方式的不同,有矿石氧化法、氧气氧化法和矿、氧综合氧化法及氩氧混吹法。

The determination of smelting method mainly depends on the composition of furnace charge and the quality requirements for the finished steel. Here we briefly introduce several smelting methods:

冶炼方法的确定主要取决于炉料的组成以及对成品钢的质量要求,下面简要介绍几种冶炼方法:

(1) Oxidation. Oxidation smelting is characterized by oxidation period. In the smelting process, oxidants are used to oxidize Si, Mn, P and other over specification elements and other impurities in the molten steel. Therefore, this method is widely used because it can smelt high quality steel with coarse materials. The disadvantage is that the smelting time is long and the easily oxidized elements are burnt.

(1) 氧化法。氧化法冶炼的特点是有氧化期,在冶炼过程中采用氧化剂用来氧化钢液中的 Si、Mn、P 等超规格的元素及其他杂质。因此,该法虽是采用粗料却能冶炼出高级优质钢,所以应用极为广泛。缺点是冶炼时间长,易氧化元素烧损大。

(2) Non oxidation method. The feature of non oxidation smelting is that there is no oxidation period. Generally, all refined materials are used, such as the returned scrap and mild steel of this steel or similar steel. The lower the content of phosphorus and other impurities, the better. The content of alloying elements added shall enter or be close to the middle or lower limit of finished steel specification. Non oxidation smelting can recover a lot of precious alloy elements and shorten smelting time. In the absence of this steel or similar steel to return to scrap, ferroalloy can be added to the furnace charge. This smelting method is also called loading method, which is indicated by the word 'in'. It is mainly used for smelting high alloy steel and other steel grades.

If no other effective measures are taken in non oxidation smelting, the hydrogen and nitrogen content in the finished steel is easy to be on the high side. In order to eliminate this disadvantage, a return oxygen blowing method has been developed.

(2) 不氧化法。不氧化法冶炼的特点是没有氧化期,一般全用精料,如本钢种或类似本钢种返回废钢以及软钢等,要求磷及其他杂质含量越低越好,配入的合金元素含量应进入或接近于成品钢规格的中限或下限。不氧化法冶炼可回收大量贵重合金元素和缩短冶炼时间。在缺少本钢种或类似本钢种返回废钢时,炉料中可配入铁合金,这种冶炼方法又称为装入法,用"入"字表示,多用于冶炼高合金钢等钢种上。

不氧化法冶炼如果不采取其他有效措施相配合,则成品钢中的氢、氮含量容易偏高。为了消除这种缺点,从而出现了返回吹氧法。

(3) Return to oxygen blowing. Return oxygen blowing method is called back blowing method for short, which is represented by the word 'return'. This method is mainly used to return scrap steel and use oxygen for slight oxidation boiling in smelting process. It can not only recover valuable alloy elements, but also reduce the content of hydrogen, nitrogen and other impurities in steel. Therefore, this method is often used to smelt chromium nickel tungsten or chromium nickel stainless steel.

（3）返回吹氧法。返回吹氧法简称返吹法，用"返"字表示。该法主要使用返回废钢并在冶炼过程中用氧气进行稍许的氧化沸腾，既有利于回收贵重的合金元素，又能降低钢中氢、氮及其他杂质的含量。因此，该法多用于冶炼铬镍钨或铬镍不锈钢等钢种。

(4) Argon oxygen mixed blowing method. After the furnace charge is fully melted, the mixed argon and oxygen gas are blown in proportion from the furnace door or from the furnace bottom, which is equivalent to one electric furnace and one AOD refining furnace. This method is mainly used in the smelting of stainless steel, characterized by high recovery of chromium, low cost, flexible and simple operation, and good quality of steel.

（4）氩氧混吹法。炉料全熔后，按比例将混合好的氩、氧气体从炉门或从炉底吹入，即相当于一台电炉又带一台 AOD 精炼炉。该法主要用于不锈钢的冶炼上，特点是铬的回收率高，成本低，操作灵活简便，且钢的质量好。

1.1.2 Principle and Task of EAF Steelmaking
1.1.2 电弧炉炼钢的原理及任务

(1) Using scrap steel as solid material and alternating current or direct current as power source, the furnace charge is heated by high temperature arc between charge and electrode to make it melt and heat up;

(2) Carry out oxidation and reduction refining in the furnace to remove harmful elements, impurities and gases in the steel;

(3) Adjust the chemical composition to the specification range;

(4) So that the temperature of liquid steel at tapping can meet the casting requirements.

（1）以废钢铁为固体料，以交流电或直流电为电源，利用炉料与电极间产生电弧的高温加热炉料，使其熔化及升温；

（2）在炉内进行氧化、还原精炼，去除钢中有害元素、杂质及气体；

（3）调整化学成分到规格范围；

（4）使钢液在出钢时的温度满足浇铸要求。

Exercises

(1) What are the main characteristics of EAF steelmaking?

(2) What are the characteristics of EAF steelmaking by oxidation process?

(3) What is the main task of EAF steelmaking?

思考题

（1）电炉炼钢的主要特征是什么？

(2) 电炉炼钢氧化法冶炼特点是什么?
(3) 电弧炉炼钢冶炼主要任务是什么?

Task 1.2　Cognition of Smelting Principle of Electric Furnace
任务1.2　电炉冶炼原理认知

Mission objectives
任务目标

(1) Master the decarbonization reaction principle of electric furnace.
(1) 掌握电炉脱碳反应原理。
(2) Master the principle of dephosphorization reaction in electric furnace.
(2) 掌握电炉脱磷反应原理。
(3) Master the desulfurization reaction principle of electric furnace.
(3) 掌握电炉脱硫反应原理。

1.2.1　Decarbonization
1.2.1　脱碳

(1) Ore decarburization. Decarbonization reaction is shown in Figure 1-1. When the temperature of molten steel is 1480~1520℃, the reaction of using ore to remove C is as follows:
(1) 矿石脱碳。脱碳反应如图1-1所示。在钢液温度为1480~1520℃时利用矿石脱碳的反应为:

$$(Fe_2O_3) + [Fe] = 3(FeO)$$

(2) Oxygen blowing and decarburization.
Direct decarburization:
(2) 吹氧脱碳。
直接脱碳反应:

$$2[C] + \{O_2\} = 2\{CO\}$$

Indirect decarburization:
间接脱碳反应:

$$2Fe + \{O_2\} = 2(FeO)$$
$$(FeO) = [O] + Fe$$
$$[C] + [O] = \{CO\}$$

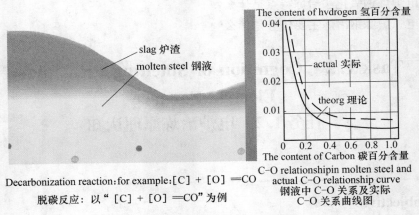

Decarbonization reaction:for example:[C] + [O] ═CO
脱碳反应: 以"[C] + [O] ═CO"为例

Figure 1-1 Decarbonization reaction
图 1-1 脱碳反应

1.2.2 Dephosphorization
1.2.2 脱磷

Dep reaction of alkaline slag:
碱性渣脱磷反应:

$$2[P] + 5[O] + 3(CaO) = (3CaO \cdot P_2O_5)$$

Dep condition: High alkalinity, high oxidation, low temperature and large slag volume.
脱磷条件: 高碱度、高氧化性、低温和大渣量。

1.2.3 Desulfurization
1.2.3 脱硫

Desulfurization reaction:
脱硫反应式:

$$[FeS] + (CaO) = (CaS) + (FeO), \Delta H > 0$$

The reaction is carried out on the slag steel interface, which is an endothermic reaction. The favorable conditions for the reaction are:

(1) High basicity, high basicity slag, increase calcium oxide in slag;

(2) Strong reducing gas (or low oxidation), reducing slag, reduce iron oxide in slag;

(3) At the same time, high temperature can improve the fluidity of slag;

(4) Large amount of slag (appropriately large), sufficient stirring to increase slag steel contact.

该反应是在渣-钢界面上进行的,为吸热反应。
有利于该反应发生的条件是:

(1) 高碱度,造高碱度渣,增加渣中氧化钙;

(2) 强还原气氛(或低氧化性),造还原性渣,减少渣中的氧化铁;

（3）高温，同时高温改善渣的流动性；
（4）大渣量（适当大），充分搅拌增加渣-钢接触。

Exercises

（1）What are the reaction formulas of ore decarburization and oxygen blowing decarburization?

（2）What are the thermodynamic conditions of dephosphorization?

（3）What are the conditions conducive to desulfurization reaction?

思考题

（1）矿石脱碳与吹氧脱碳的反应式分别是什么？

（2）脱磷反应的热力学条件是什么？

（3）有利于脱硫反应发生的条件是什么？

Task 1.3　Cognition of Raw Materials for Electric Furnace Smelting
任务1.3　电炉冶炼原材料认知

Mission objectives

任务目标

（1）Understand the types of raw materials for EAF steelmaking.

（1）了解电炉炼钢原材料种类。

（2）Master the role of various raw materials in smelting process.

（2）掌握各种原材料在冶炼过程中的作用。

The management and use of raw materials for steelmaking is an important part of the production of EAF steelmaking, which directly affects the quality, output, consumption and other economic and technical indicators of steel. Therefore, all kinds of raw materials for steelmaking should meet the relevant national standards. The charge management shall be strict and meticulous, classified storage, superior material not inferior, coarse material fine processing.

炼钢用原材料的管理与使用是电炉炼钢生产中的重要组成部分，直接影响钢的质量、产量、消耗等经济技术指标。因此炼钢用各种原材料应符合有关国家标准。炉料管理应做到严格细致，分类保管，优料不劣用，粗料细加工。

1.3.1　Steel Scrap
1.3.1　废钢

（1）Scrap source：

1) Common scrap. It has a wide range of sources and complex composition and specifications.

2) Return to scrap. From the smelting and processing workshop of the iron and steel plant.

（1）废钢来源：

1）普通废钢。来源广泛，成分和规格较复杂。

2）返回废钢。来自钢铁厂的冶炼和加工车间。

（2）Requirements for scrap：

1) The surface of scrap steel is clean and less rusty.

2) The scrap shall not be mixed with non-ferrous metals such as copper, lead, zinc, tin, antimony and arsenic.

3) No explosives, inflammables, closed utensils and drugs shall be mixed in the scrap.

4) Scrap should have clear chemical composition.

5) Scrap steel shall have proper block size and overall dimension, see Table 1-1.

（2）对废钢的要求：

1）废钢表面清洁、少锈。

2）废钢中不得混有铜、铅、锌、锡、锑、砷等有色金属。

3）废钢中不得混有爆炸物、易燃物、封闭器皿和毒品。

4）废钢要有明确的化学成分。

5）废钢要有合适的块度和外形尺寸，见表1-1。

Table 1-1 Boundary dimension and weight requirements of all kinds of scrap

表1-1 各类废钢外形尺寸及质量要求

Type 类型	Maximum size/mm 最大尺寸/mm	Maximum area/mm^2 最大面积/mm^2	Maximum weight/kg 最大质量/kg
Large scrap 大型废钢	Long 1200 长1200	800×600	800
Medium scrap 中型废钢	Long 800 长800	500×300	200
Small scrap 小型废钢	Long 400 长400	200×200	50
Note surplus, package bottom 注余、包底	High 400 高400	$\phi1200$	1500

1.3.2 Pig Iron
1.3.2 生铁

It is generally used to increase the carbon content of furnace charge or replace part of scrap steel. Usually the dosage is 10%~20%. Pig iron for steelmaking must be stacked according to the batch number. Pig iron for carburizing should be low rust, dry and low phosphorus pig iron.

生铁一般用于提高炉料的配碳量或代替一部分废钢。通常配入量为10%~20%。炼钢用生铁必须按批号堆放。增碳用生铁应是表面少锈、干燥的低磷生铁。

1.3.3 Oxidant
1.3.3 氧化剂

(1) Iron ore, scale. It is required that the iron content of iron ore is high and the ratio is significant.

(2) Oxygen. It is the most important oxidant. Oxygen content is required to be not less than 99%, water content is not more than $3g/m^3$, oxygen pressure in melting period is 0.3~0.7MPa, and oxygen pressure in oxidation period is 0.7~1.2MPa.

(1) 铁矿石、氧化铁皮。要求铁矿石的含铁量高,相对密度大。

(2) 氧气。氧气是最主要的氧化剂。要求含O_2不小于99%,水分不大于$3g/m^3$,熔化期氧压为0.3~0.7MPa,氧化期氧压为0.7~1.2MPa。

1.3.4 Deoxidizer and Ferroalloy
1.3.4 脱氧剂和铁合金

In the process of electric furnace smelting, the commonly used deoxidizer components include ferrosilicon, ferromanganese, calcium silicon, aluminum, carbon, etc. The commonly used alloys are ferrosilicon, ferromanganese, silicomanganese, ferrochrome, silicochromium, ferromolybdenum, ferrovanadium, ferrotitanium, etc.

在电炉冶炼过程中,常用的脱氧剂成分包括硅铁、锰铁、硅钙、铝、碳等。常用的合金有硅铁、锰铁、锰硅、铬铁、硅铬、钼铁、钒铁、钛铁等。

The requirements for deoxidizer and ferroalloy are as follows:

(1) The content of useful elements in Deoxidizers and ferroalloys should be high to reduce the heat consumption during melting.

(2) The block size should be appropriate, the block size that is difficult to melt should be smaller, and the block size that is easy to melt should be larger.

(3) The content of nonmetallic inclusions, P, S, gas and slag should be as low as possible.

(4) Try to use high carbon ferroalloy first, then medium and low carbon ferroalloy within the scope of permission.

(5) The iron alloy which is difficult to oxidize and melt should be heated before use.

对脱氧剂和铁合金要求是:

(1) 脱氧剂和铁合金中的有用元素含量要高,以减少熔化时的热能消耗。

(2) 块度大小应适当,难熔化的块度要小些,易熔的可大些。

(3) 非金属夹杂物、P、S和气体以及炉渣等杂质含量应尽量低些。

(4) 在许可范围内尽量先使用高碳铁合金,然后使用中、低碳铁合金。

(5) 不易氧化且难熔化的铁合金在使用前应加热。

1.3.5 Slagging Material
1.3.5 造渣材料

(1) Lime. Requirements: $w(CaO) \geqslant 85\%$, $w(SiO_2) < 2\%$, $w(S) < 0.5\%$, active lime should be used.

(2) Fluorite. Requirements: $w(CaF_2) = 85\% \sim 95\%$, which should be used after drying at low temperature of $100 \sim 200℃$.

(3) Clay brick. $w(SiO_2) \approx 60\%$, $w(Al_2O_3) \approx 35\%$, improving slag fluidity.

(1) 石灰。要求: $w(CaO) \geqslant 85\%$, $w(SiO_2) < 2\%$, $w(S) < 0.5\%$, 宜采用活性石灰。

(2) 萤石。要求: $w(CaF_2) = 85\% \sim 95\%$, 应在 $100 \sim 200℃$ 的低温干燥后使用。

(3) 黏土砖块。$w(SiO_2) \approx 60\%$, $w(Al_2O_3) \approx 35\%$, 改善炉渣流动性。

1.3.6 Carburizer
1.3.6 增碳剂

(1) Toner. It can remove iron oxide from slag and reduce oxygen partial pressure in gas phase. In the deoxidization process, it is inevitable to increase part of the carbon content.

(2) Electrode powder (broken electrode block). It is the best carburizing material with a large ratio (broken electrode block), low S and P, and low ash content.

(1) 碳粉。碳粉既是增碳剂也是还原剂,可脱除炉渣中的氧化铁和降低气相中的氧气分压力。在脱氧过程中,不可避免地也要增加部分碳含量。

(2) 电极粉(碎电极块)。电极粉是最好的增碳材料,其相对密度大(碎电极块),S、P含量低,灰分少。

1.3.7 Electrode
1.3.7 电极

(1) Performance of graphite electrode:

1) After heating, graphite is directly sublimated from the original state to the gas state, and the sublimation temperature is as high as $3800℃$.

2) Different from most of the materials, the mechanical strength of graphite increases with the increase of temperature, and the strength of graphite is 1.6 times of that at $2000℃$.

3) The dependence of the resistance coefficient of graphite on the temperature variation is complex. The resistance coefficient at $1400℃$ is the same as that at room temperature, but the resistance coefficient of metal always increases with the increase of temperature.

4) Graphite has good thermal conductivity and low coefficient of thermal expansion, which makes graphite have good thermal shock resistance and reduces the thermal stress in the electrode.

5) When the surface temperature of graphite is higher than $400℃$, it will combine with oxygen. The amount of oxidation is related to the oxygen content, gas flow rate and exposure

time. When the temperature is higher than 600℃, the oxidation process is intense.

6) Graphite is easy to process, the screw joints at both ends of the electrode have high processing accuracy, good contact at the joints, good mechanical stress distribution, and the price is not expensive.

（1）石墨电极性能：

1）石墨加热后，直接由故态升华为气态，升华温度高达3800℃。

2）它与大部分材料不同，在温度上升时，其机械强度上升，在2000℃时石墨强度是室温下的1.6倍。

3）石墨的电阻系数和温度变动的相依关系较复杂。在1400℃下的电阻系数和室温下的相同，而金属的电阻系数却总是随温度升高而增大的。

4）石墨的导热性能好而热膨胀系数较低，使石墨抗热震性能较好，降低了电极中的热应力。

5）在石墨表面温度大于400℃后会和氧气结合。氧化量与气体中的氧含量、气体流速和暴露时间有关。在温度大于600℃后，氧化过程激烈。

6）石墨易于加工，电极两端螺纹接头有较高的加工精度，接头处接触良好，机械应力分布较好，且价格不贵。

（2）Consumption of graphite electrode.

1) Electrode end consumption:

① Sublimation.

② Spalling due to thermal stress.

③ Melting.

2) Electrode side consumption.

3) Electrode break——high position break.

4) Residual consumption——low level fracture.

（2）石墨电极的消耗。

1）电极端面消耗：

①升华。

②热应力导致剥落。

③熔解。

2）电极侧面消耗。

3）电极折断——高位断裂。

4）残端消耗——低位断裂。

（3）Measures to reduce electrode consumption:

1) Reduce the breakage and breakage of electrode caused by mechanical and electromagnetic force.

2) The electrodes shall be stored in a dry place to prevent moisture.

3) Reduce the electric loss of electrode joint.

4) Reduce the oxidation consumption around the electrode.

5) Reduce the consumption of electrode end.

6) Improving the sensitivity of electrode regulator's automatic rising and falling is beneficial to reducing electrode consumption.

7) Strengthen management and strictly implement key operation systems such as power distribution, batching and charging to reduce abnormal consumption.

（3）降低电极消耗的措施：

1）减少由机械外力和电磁力引起电极折断和破损。

2）电极应存放在干燥处，严防受潮。

3）减少电极接头的电损失。

4）减少电极周界的氧化消耗。

5）降低电极端部的消耗。

6）提高电极调整器自动升降的灵敏度，有利于降低电极消耗。

7）强化管理，严格执行配电、配料、装料等关键操作制度，以减少非正常消耗。

Exercises

（1）What are the requirements for scrap entering the electric furnace?

（2）What are the main deoxidizers for electric furnace smelting?

（3）What are the requirements for deoxidizer and ferroalloy?

（4）What are the slag making materials for electric furnace smelting?

（5）What are the factors influencing the consumption of graphite electrode?

思考题

（1）入电炉废钢有哪些要求？

（2）电炉冶炼的脱氧剂主要有哪些？

（3）对脱氧剂和铁合金的要求是什么？

（4）电炉冶炼造渣材料包括哪些？

（5）影响石墨电极消耗的因素包括什么？

Project 2　Production Preparation
项目2　生产准备

Task 2.1　Mechanical Equipment Preparation
任务2.1　机械设备准备

Mission objectives
任务目标

(1) Master the composition of mechanical equipment.
(1) 掌握机械设备的构成。
(2) Master the requirements of the production process on mechanical equipment.
(2) 掌握生产工艺对机械设备的要求。

2.1.1　Equipment Composition
2.1.1　设备构成

The structure of EAF includes furnace metal components, electrode holders, electrode lifting device, furnace tilting device, roof lifting and rotating device, as shown in Figure 2-1.

电弧炉由炉体金属构件、电极夹持器、电极升降装置、炉体倾动装置、炉盖提升和旋转装置等构成,如图2-1所示。

2.1.2　Furnace Body
2.1.2　炉体

Furnace body is the most important equipment of EAF, which is used to melt charge and carry out various metallurgical reactions. The main body is composed of two parts: metal components and refractory linings.

炉体是电炉最主要的装置,用来熔化炉料和进行各种冶金反应。主体由金属构件和耐火材料砌成的炉衬两部分组成。

Metal components of furnace body: furnace shell, furnace door, tapping trough, roof ring and electrode sealing ring.

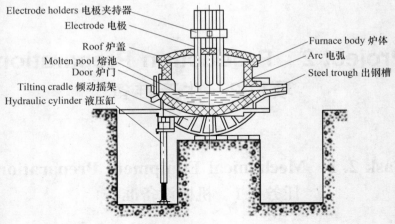

Figure 2-1　Structure of EAF
图 2-1　电弧炉结构图

炉体的金属构件有炉壳、炉门、出钢槽、炉盖圈和电极密封圈。

The furnace shell is welded with steel plate, and its upper part is provided with a reinforcing ring.

炉壳是用钢板焊接而成的,它的上部设有加固圈。

The structure of furnace shell includes furnace shell, furnace shell bottom and upper reinforcing ring.

炉壳的结构包括炉身壳、炉壳底和上部加固圈。

2.1.2.1　Furnace Shell
2.1.2.1　炉壳

(1) Body of furnace shell. Requirements: in the process of steelmaking, it is required to bear the weight of furnace lining and charge, as well as the impact force during charging and the thermal stress when the furnace lining is heated. Therefore, the furnace shell is required to have sufficient strength and rigidity. The standby furnace shell is used to ensure normal production and improve the furnace operation rate.

(1) 炉身壳。要求:炼钢过程中除要承受炉衬和炉料的重量外,还要抵抗装料时的冲击力,还受到炉衬被加热时的热应力。因此要求炉壳具有足够的强度和刚度。采用备用炉壳保证正常生产,提高炉子作业率。

The thickness of steel plate of furnace shell is related to the diameter of furnace shell. According to experience, it is about 1/200 of the diameter of furnace shell, and the thickness is usually 12~30mm. In order to accelerate the drying of the new furnace lining, a small hole with a diameter of about 20mm is drilled in the furnace shell, and the hole distance is about 400~500mm.

炉壳钢板厚度与炉壳直径大小有关,根据经验大约为炉壳直径的 1/200,通常厚度为

12~30mm。为加速新炉衬的干燥，炉壳上钻有直径约20mm的小孔，孔距约400~500mm。

The shell of the furnace body is made into cylinder shape to reduce the heat dissipation area and heat loss. The notch around the furnace door and tapping hole needs to be reinforced with steel plate.

炉身壳做成圆筒形，减少散热面积和热损失，炉门和出钢口四周的切口部分需要用钢板加固。

(2) Bottom of furnace shell. The shape of the bottom of the furnace shell is spherical, frusto-conical and flat bottom.

(2) 炉壳底。炉壳底部形状有球形、截头圆锥形和平底形三种。

The spherical bottom has the strongest rigidity, the small dead angle, and the smallest lining volume, but the large-diameter spherical bottom is difficult to form, so it is mostly used for small and medium-sized furnaces. The frusto-conical furnace shell bottom is slightly weaker than the spherical bottom. It requires a little more lining material, but it is easy to manufacture, so it is often used. The flat bottom furnace shell has the worst robustness and the largest lining volume, but it is simple to manufacture, so it is mostly used on large electric furnaces.

球形底坚固性最大，死角小，炉衬体积最小，但直径大的球形底成形困难，故多用于中小型炉子。截头圆锥形炉壳底，其坚固性较球形底略差，所需的炉衬材料稍多，但制造容易，故常采用。平底炉壳坚固性最差，炉衬体积最大，但制造简单，故多用于大型电炉上。

(3) Reinforcing ring on top edge of furnace shell. It is welded with steel plate or section steel, and large EAF adopts middle water cooling reinforcing ring to increase the rigidity of furnace shell, prevent deformation due to heating and ensure close contact between furnace shell and roof. In recent years, the upper part of the slag line has been cooled by water, and a water-cooled furnace shell with an interlayer is used. Large furnaces also use ribs. A sand sealing groove is left on the upper part of the reinforcement ring, so that the roof ring is inserted into the sand groove for easy sealing.

(3) 炉壳上沿的加固圈。用钢板或型钢焊成，大型电炉采用中间通水冷却的加固圈，增加炉壳刚度，防止受热而变形，保证炉壳与炉盖接触紧密。近年来渣线以上部分都通水冷却，使用带夹层的水冷炉壳。大型炉子还采用加强筋。在加固圈的上部留有一个砂封槽，使炉盖圈插入沙槽内便于密封。

2.1.2.2 Door
2.1.2.2 炉门

The door includes furnace door cover, frame, sill and lifting mechanism, as shown in Figure 2-2.

炉门包括炉门盖、炉门框、炉门槛和炉门升降机构，如图2-2所示。

Requirements: tight structure, simple and flexible lifting, strong and durable, easy to disassemble.

Figure 2-2　Door
图 2-2　炉门

要求：结构严密，升降简便灵活，坚固耐用，便于拆装。

The furnace door cover is welded with steel plate and adopted hollow water cooling type. The upper part of the door frame is embedded in the furnace wall to support the furnace wall above the furnace door. The front wall of the furnace door frame is inclined, with an angle of 8°~12° with the vertical line, so as to ensure that the furnace door cover and the furnace door frame can be compressed and well sealed, so as to reduce the heat loss and maintain the atmosphere in the furnace.

炉门盖用钢板焊成，采用空心水冷式，炉门框上部嵌入炉墙内，支撑炉门上部的炉墙，炉门框前壁做成倾斜的，和垂线成 8°~12° 的夹角，保证炉门盖和炉门框之间能压紧，密封良好，减少热量损失和保持炉内气氛。

Most use hollow water-cooled furnace doors. There are four lifting mechanisms: manual, pneumatic, hydraulic transmission and electric. The lifting mechanism of the furnace door can be manually operated on small furnaces, and the large and medium furnaces are pneumatically or hydraulically driven. Its lifting should be stable, and the furnace door can stay at any position as required.

大多数采用空心水冷的炉门。升降机构有手动、气动、液压传动和电动四种。炉门的升降机构在小炉子上可用手动，大中型炉子采用气动或液压传动。其升降要平稳，并可根据需要而使炉门停留在任何位置上。

2.1.2.3　Tapping Trough and Eccentric Bottom Tapping
2.1.2.3　出钢槽和偏心底出钢

The steel outlet groove is welded by steel plate and angle steel and fixed on the furnace shell. There are large refractory bricks in the groove. At present, EBT is widely used.

出钢槽由钢板和角钢焊成，固定在炉壳上，槽内有大块耐火砖。目前多是偏心底出钢。

In January 1983, the first EAF with EBT was put into operation in Denmark special steel works, which was transformed from 110t.

1983年1月,第一台偏心底出钢电弧炉在丹麦特殊钢厂投产,是110t改造炉。

In June 1987, the first eccentric bottom tapping electric arc furnace was built in Shanggang No. 5 plant.

1987年6月,我国第一台偏心底出钢电弧炉在上钢五厂建成。

EBT principle: install the tapping box protruding the furnace shell at the original tapping port side to replace the original tapping slot. The refractory materials are laid inside to form a small melting pool, which is connected with the large melting pool at the bottom of the original furnace and has a smooth transition. The tapping hole is vertically opened at the bottom of the small melting pool of the tapping box. The upper water-cooled roof plate is used to close the small melting pool and clean and maintain the tapping hole.

EBT原理:原出钢口侧安装突出炉壳的出钢箱以取代原来的出钢槽。内部砌筑耐火材料,形成一个小熔池,与原炉底大熔池连通且圆滑过渡。出钢口垂直地开在出钢箱小熔池的底部。上部设水冷盖板,以封闭小熔池及清理与维护出钢口。

Double layer structure: the outer layer is square block, the inner layer is sleeve brick, and the layer to layer is filled with magnesia refractory to facilitate the replacement of sleeve brick. The opening and closing can be completed by opening and closing the swinging roof plate. Tail brick is cooled with water. Roof plate and tail brick are made of graphite, which are used to prevent high temperature deformation and protect furnace. Before charging, it is necessary to close the roof plate of the tapping hole and fill the tapping hole with $MgO-SiO_2$ mixed powder containing 10% Fe_2O_3 to block the tapping hole. EBT mechanism is shown in Figure 2-3.

双层结构:外层为方形座砖,内层为袖砖,层与层间用镁质耐火材料填充,以便于袖砖的更换。开闭可通过开闭摆动式盖板完成。尾砖用水冷却,盖板与尾砖由石墨构成,以防高温变形与保护炉子。装料前,关闭出钢口盖板,在出钢口内填入10% Fe_2O_3(质量分数)的$MgO-SiO_2$混合粉料,堵塞出钢口。偏心底出钢机构如图2-3所示。

The maximum inclination of the furnace body is 12°~15°, which is convenient for the maintenance of the tapping hole, and the steel slag can be cleaned and retained.

最大倾角12°~15°,以便于出钢口的维护,钢渣既能出得净又留得住。

The opening and closing of the steel outlet is realized by the cylinder driving the connecting rod and crank mechanism to control the movement of the skateboard. Including: cylinder, adjusting bolt, connecting rod, crank, large skateboard, small skateboard, skateboard base and skateboard guide wheel.

出钢口的开启和关闭是由气缸带动连杆、曲柄机构来控制滑板动作实现的,由气缸、调整螺栓、连杆、曲柄、大滑板、小滑板、滑板底座及滑板导轮组成。

Advantage:

(1) Complete slag free tapping and slag retention operation, up to more than 95%.

(2) Increase the water-cooling area from 70% to 87%~90%, and increase the life of EAF.

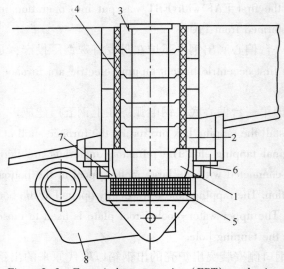

Figure 2-3 Ecentric bottom tapping (EBT) mechanism
1—bottom brick; 2—steel brick; 3—steel pipe; 4—mixed plastic;
5—graphite plate; 6—water cooling; 7—bottom ring; 8—cover plate

图 2-3 偏心底出钢 (EBT) 机构
1—底砖；2—出钢砖；3—出钢管；4—混合可塑料；
5—石墨板；6—水冷；7—底环；8—盖板

(3) The angle of inclination is reduced, the length of short network is shortened, the active power and power factor in the furnace are increased, the smelting time is shortened, and the power consumption is reduced by 15%~30%. Simplify furnace design.

(4) Reduce the tapping time by 75% and the tapping temperature by 30℃.

(5) The tapping flow is vertical and concentrated without dispersion. In the process of tapping, the secondary oxidation and gas absorption of the steel flow are reduced, the tapping time is short, and the contents of hydrogen, oxygen, nitrogen and inclusions in the steel are reduced. Roof and argon protection.

优点：

(1) 彻底无渣出钢和留钢留渣操作，可达95%以上；

(2) 增加水冷面积，70%到87%~90%，提高电炉寿命。

(3) 倾角减小，缩短短网长度，从而提高输入炉内的有功功率和功率因数，缩短冶炼时间，降低电耗15%~30%，简化炉子设计。

(4) 缩短出钢时间75%，降低出钢温度30℃。

(5) 出钢钢流短而垂直，且集中无分散。减轻出钢过程中钢流的二次氧化及吸气，出钢时间短，钢中氢、氧和氮及夹杂物的含量均有所减少。加盖及氩气保护。

2.1.2.4 Roof Ring

2.1.2.4 炉盖圈

The role of roof ring is to keep the correct shape of furnace top and to put the whole mass of roof on the furnace shell. Roof ring is welded with steel plate or section steel to form an inclined inner wall instead of arch foot brick, with an inclined angle of about 22°~23°. In order to prevent deformation, cold water is generally used. There must be a good seal between the roof ring and the furnace shell, so the lower part of the outer edge of the furnace shell ring is provided with a knife edge, so that the roof ring can be well inserted into the sand sealing groove of the reinforcement ring. A lifting ring and a baffle are welded on the roof ring, so as to lift the roof and prevent the roof from moving relative to the furnace body when tilting the furnace. The section shape of roof circle is shown in Figure 2-4.

炉盖圈的作用是保持炉顶的正确形状以及把炉盖的全部质量落在炉壳上。炉盖圈用钢板或型钢焊成,做成倾斜形内壁以代替拱脚砖,倾斜角约22°~23°,为了防止变形,一般采用冷水。炉盖圈和炉壳之间必须有良好的密封,因此炉壳圈外沿下部设有刀口,使炉盖圈能很好地插入到加固圈的砂封槽内。炉盖圈上焊有吊环和挡板,以便吊装炉盖和防止倾炉时炉盖相对于炉体移动。炉盖圈截面形状如图2-4所示。

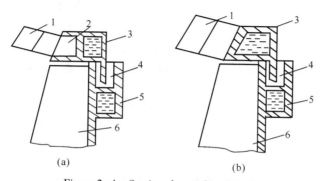

Figure 2-4 Section shape of roof circle
(a) vertical roof circle; (b) inclined roof circle
1—roof; 2—arch foot brick; 3—roof circle; 4—sand trough;
5—water-cooling reinforcing circle; 6—furnace wall

图 2-4 炉盖圈截面形状
(a) 垂直形炉盖圈; (b) 倾斜形炉盖圈
1—炉盖; 2—拱脚砖; 3—炉盖圈; 4—沙槽;
5—水冷加固圈; 6—炉墙

2.1.2.5 Electrode Sealing Ring
2.1.2.5 电极密封圈

In order to make the electrode free to rise and fall, it is required that the diameter of the electrode hole should be 40~50mm larger than the electrode diameter. The high-temperature gas in the furnace will continue to escape from the gap of the electrode hole, and the cold air will be continuously sucked from the gap of the furnace door. The temperature of the upper electrode rises violently and the electrode becomes thin and easily broken. The installation of electrode sealing ring (see Figure 2-5) can improve the working conditions of the electrode chuck and prevent the furnace gas from escaping. At the same time, it can also cool the roof and extend the life of the roof.

为了使电极可自由升降,要求电极孔的直径应比电极直径大 40~50mm。炉内的高温气体将会不断地从电极孔缝隙中逸出,而冷空气则不断地从炉门缝隙中吸入,这样不但使热损失增加,不利于维护炉内还原性气氛,而且使炉盖上部的电极温度升高氧化激烈,电极变细而易折断。设置电极密封圈(见图 2-5),可以改善电极夹头的工作条件,防止炉气外逸,同时也可冷却炉盖,延长炉盖寿命。

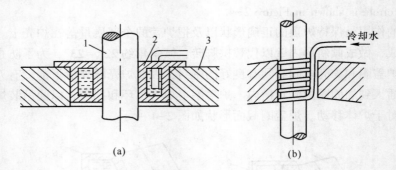

Figure 2-5　Electrode sealing ring
(a) annular water tank type electrode seal ring; (b) serpentine tube type electrode seal ring
1—electrode; 2—electrode sealing ring; 3—roof
图 2-5　电极密封圈
(a) 环形水箱式电极密封圈;(b) 蛇形管式电极密封圈
1—电极;2—电极密封圈;3—炉盖

The electrode sealing ring of the brick roof has two types: serpentine water pipe type and annular water tank type. The sealing and sealing effect of the annular water tank sealing ring is better, and it is one widely used at present.

砖砌炉盖的电极密封圈,有蛇形水管式和环形水箱式两种。环形水箱密封圈冷却和密封效果均较好,目前已被广泛采用。

2.1.3　Electrode Arms and Electrode Lifting Device
2.1.3　电极横臂及电极升降装置

2.1.3.1　Electrode Arms
2.1.3.1　电极横臂

The electric arc furnace has three electrodes, each of which is supported by electrode arms. Electrode arms include: the main body of the arm, the holder, and the conductive copper tube that transmits large current. The main function of electrode arms is to support and hold the electrodes. Its structure is a hollow box structure welded by steel pipes and steel plates. Because electrode arms work in a strong magnetic field and electric field environment, eddy currents are generated during work and heat is generated, so internal reinforcement and water cooling are used. Arms are also provided with conductive copper tiles connected to the conductive copper pipes. The copper tiles and the copper pipes are filled with cooling water to cool the conductive copper pipes and the copper tiles. The conductive copper tube and the electrode chuck must maintain good insulation from the uncharged mechanical structure in the cross arm to prevent the furnace body from being charged. The latest development of electrode arms manufacturing technology is the use of copper steel composite plates or aluminum to make conductive crossarms, which not only reduces impedance and saves electricity, but also reduces the weight of electrode arms and reduces the maintenance work on copper pipes.

电弧炉有三支电极,每支电极都靠电极横臂支撑。电极横臂包括横臂主体、夹持器、传输大电流的导电铜管。电极横臂的主要作用是用来支撑、把持电极。其结构是钢管和钢板焊接成中空的箱型结构。由于电极横臂工作在强大的磁场和电场环境内,工作时会产生涡流而发热,故内部采用加强筋和水冷却。横臂上还设置了与导电铜管相连的导电铜瓦,铜瓦和铜管内部通以冷却水,对导电铜管和铜瓦进行冷却。导电铜管和电极夹头必须与横臂中不带电的机械结构部分保持良好的绝缘,以防止炉体带电。电极横臂制造技术的最新发展是采用铜钢复合板或铝制造导电横臂,不仅可减少阻抗而节约电能,而且还减轻电极横臂的重量,减少对铜管的维护工作量。

2.1.3.2 Electrode Holder
2.1.3.2 电极夹持器

The role of the electrode holders is to clamp or loosen the electrodes, and then transfer the current and voltage from the short net work to the electrodes. The electrode holders are composed of front chucks, conductive copper tiles, connecting rods, stack springs and hydraulic cylinders.

电极夹持器的作用是夹紧或松放电极,并将短网传送过来的电流和电压再传送到电极上。电极夹持器由前夹头、导电铜瓦、连杆、叠簧和液压缸等几部分组成。

The tension of the spring clamps the electrode, and the force generated by the hydraulic cylinder compresses the spring in the opposite direction to loosen the electrode. It is characterized by easy operation and low labor intensity.

弹簧的张力夹紧电极,利用液压缸产生的力将弹簧反方向压缩而松开电极。它的特点是操作简便、劳动强度小。

The chuck can be made of copper or steel. The copper chuck has good conductivity and small resistance, but the mechanical strength is poor, and the expansion coefficient is large; the steel chuck has high strength, but the resistance is large, and the electrical loss increases. If the chuck is made of non-magnetic steel, the electromagnetic loss can be reduced. Water cooling is needed in the middle of the chuck to ensure sufficient strength, reduce expansion, and also play a role in reducing oxidation and lowering resistance. The electrode chuck is fixed on the cross arm, and the front chuck has no conductive effect.

夹头可用铜或钢制成。铜制夹头的导电性能好,电阻小,但机械强度差,膨胀系数大;钢制夹头的强度高,但电阻大,电损耗增加,如采用无磁性钢制作夹头,则可减少电磁损失。夹头中间需通水冷却，以保证足够的强度，减少膨胀，还可起到减少氧化、降低电阻的作用。电极夹头固定在横臂上，前夹头没有导电作用。

The electric copper tile is connected with the conductive copper tube and the cross arm, and there is an insulating material (HP-5) between the copper tile and the cross arm. The copper tile and the conductive copper tube are connected by a non-magnetic stainless steel bolt. The copper tile material is TU_2 oxygen-free copper.

电铜瓦与导电铜管和横臂相连，铜瓦与横臂间有绝缘材料（HP-5），铜瓦与导电铜管采用非磁性不锈钢螺栓连接，铜瓦材料为 TU_2 无氧铜。

There are many ways to clamp and release electrode, such as clamp, wedge, screw and pneumatic spring.

夹紧和松放电极方式很多，有钳式、楔式、螺旋压紧式和气动弹簧式等。

2.1.3.3 Mast and Guidance System
2.1.3.3 立柱及导向系统

Each column is constrained by several guide rollers, which are divided into positive guide rollers and side guide rollers. Each guide roller can be adjusted to ensure that the column is perpendicular to the plane of the cradle, and the verticality requirement is less than 0.5/1000. The verticality of the column is completed by the adjustment of the front guide roller and the side guide roller. The distance between the electrode arms is adjusted by the side guide rollers. In actual operation, the adjustment force of the guide roller should be moderate. Too tight will cause the lead screw to have a large force and shorten the life of the screw. The vibration becomes larger and it is easy to cause accidents.

每个立柱由若干个导向辊进行约束，分为正导辊和侧导辊。每个导辊都可进行调整，以保证立柱相对摇架平面的垂直，垂直度要求小于0.5/1000。其中立柱的垂直度靠正导辊和侧导辊的调整来完成，而电极横臂间的间距靠侧导辊的调整来完成。在实际操作中，导向辊的调节力度要适中，太紧会使导向辊的丝杠受力大、丝杠寿命缩短，太松则不能起到对立柱的固定作用，在生产过程中电横臂的振动变大，容易产生事故。

2.1.3.4 Electrode Lifting Device
2.1.3.4 电极升降装置

It is required that the electrode lifting device has the advantages of flexible lifting, small inertia of the system, start, quick starting and braking; lifting speed must be adjustable; rise faster and fall slower, in order to avoid frequent tripping of circuit breakers and excessive fluctuations of current and voltage, it will adversely affect the operation.

电极升降装置要求升降灵活、系统惯性小、启动、制动快；升降速度要能够调节，上升要快、下降要慢，以避免断路器的频繁跳闸和电流、电压过大波动，对操作产生不利影响。

The adjustment object of the electrode automatic adjustment system is the arc length, and its means is to rely on accurately controlling the position of the electrode movement. At present, there are many types of electrode automatic adjustment devices, but the electrode automatic adjustment process is composed of the following links:

(1) Measurement and comparison. Its function is to measure the current and voltage of the electric furnace and compare it with the given value, and then add the difference after comparison as the signal voltage to the amplifier of the system.

电极自动调节系统调节的对象是电弧长度,其手段是依靠准确地控制电极升降的位置来实现。目前电极自动调节装置类型较多,但电极自动调节过程均是由以下几个环节构成:

(1) 测量比较。其作用是测出电炉的电流和电压并与给定值进行比较,然后将比较后的差值作为信号电压加于系统的放大器。

(2) Amplification. Because the signal power output from the measurement and comparison link is usually very small, it is not enough to drive the lifting electrode mechanism, so an amplification link must be installed in the system. There are many amplifiers, and the electrode automatic adjustment device has different names and performances due to the different amplifiers used.

(2) 放大环节。由于测量比较环节所输出的信号功率通常很小,不足以推动升降电极的机构,因而在系统中必须装有放大环节。放大器有许多种,电极自动调节装置由于所用的放大器不同,而具有不同的名称和性能。

(3) Actuator. It acts according to the signal sent by the amplifier to adjust the electrode up and down.

(3) 执行机构。它根据放大器送来的信号动作,对电极进行升降调节。

(4) Type. Fixed column type (lift truck type); movable column type. Large and medium EAF adopts movable column type.

(4) 类型。固定立柱式(升降车式);活动立柱式。大中型 EAF 采用活动立柱式。

(5) Two transmission modes. Hydraulic transmission and electric transmission.

(5) 两种传动方式。液压传动和电动传动。

2.1.4　Roof Lifting System
2.1.4　炉盖提升系统

The roof lifting mechanism used in the electric arc furnace is a connecting rod mechanism, and the principle of parallel four connecting rods is used to realize the lifting and lowering of the furnace cover. At the same time, a rotation lock device for the roof is also provided in the mechanism. It mainly includes crank, connecting rod and lifting ear. The source of power for lifting is two hydraulic cylinders connected in series on the gantry, and the lowering is achieved by the weight of the roof. The connecting rod and crank are connected by the pin shaft.

电弧炉使用的炉盖提升机构为连杆机构,采用平行四连杆原理实现了炉盖的起落。同时,在机构中还设置了炉盖的旋转锁定装置,它主要包括曲柄、连杆和吊耳。提升的动力来源是安装在门型架上的两个串联的液压缸,而下降靠炉盖的自重来实现。连杆与曲柄的连接靠销轴来实现。

2.1.5 Furnace Tilting Device
2.1.5 炉体倾动装置

Requirements: it can be tilted 10°~15° to facilitate slag raking and other operations. The tilting speed shall be adjustable, and it shall be stable and reliable during tilting, so as not to overturn the furnace; the end of the steel chute shall be moved in a small horizontal direction, so as to avoid a large forward and backward moving distance of the steel barrel.

要求：可以倾动 10°~15°，以便于扒渣等操作。倾动速度应能调节，倾动时平稳可靠，不致翻倒炉子；出钢槽末端在水平方向移动要小，以免盛钢桶前后移动距离较大。

The electric furnace tilting system consists of tilting cylinder, cradle rail, cradle, and tilt lock.

电炉倾动系统由倾动缸、摇架轨道、摇架、倾动锁定组成。

The upper track of the cradle is an arc-shaped track with positioning teeth; the lower track has positioning holes, and the surface of the lower track has a certain slope, the steel exit side is 1°, and the slag exit side is 2°. The position constraint when the furnace body is tilted is achieved by positioning teeth and positioning holes. The tilt lock is achieved by tilting the lock plate on the side of the electric furnace.

摇架上轨道为带定位齿的弧形轨道；下轨道有定位孔，且下轨道表面具有一定斜度，出钢侧为 1°，出渣侧为 2°。炉体倾动时的位置约束靠定位齿和定位孔来实现。倾动的锁定靠在电炉一侧的倾动锁定插板来实现。

Type: roll and bottom tilt.
类型：侧倾和底倾两种。

2.1.6 Top Charging System
2.1.6 炉顶装料系统

The charging time can be shortened and the labor intensity can be reduced by using the top charging, and the volume of the furnace and the large charging can be fully utilized.

采用炉顶装料可缩短装料时间，减轻劳动强度，并且可以充分利用炉膛的容积和装入大块炉料。

Type: roof rotary type, furnace body outgoing type, roof outgoing type.
类型：炉盖旋转式，炉体开出式，炉盖开出式。

Roof rotating type: cantilever (top frame), electrode lifting system are installed on the suspension, roof is hung on the cantilever. Raise the electrode and roof first, then rotate the entire cantilever with roof and electrode system toward the steel outlet-transformer side by 70°~90° to expose the furnace for loading.

炉盖旋转式：悬臂架（顶架），电极升降系统都装在悬架上，炉盖吊挂在悬臂架上面。先升高电极和炉盖，然后整个悬臂架同炉盖和电极系统向出钢口—变压器一侧旋转 70°~90°，以露出炉膛进行装料。

Advantages: the mass of the rotating part is small, the mass of all the metal structures of the furnace is minimum, and the action is rapid.

优点：旋转部分质量较小，炉子全部金属结构质量最小，动作迅速。

Disadvantages: the distance between the furnace center and the transformer is large, increasing the length of the short network.

缺点：炉子中心与变压器距离较大，增大短网长度。

The driving method of the roof lifting mechanism and the furnace body opening mechanism adopts electric and hydraulic transmission. The movable workbench is an electric trolley. It is necessary to use a special material blue to charge the furnace.

炉盖提升机构和炉体开出机构的传动方式均采用电动和液压传动。活动工作台为一个电动台车。装料需要专门的料篮。

Furnace top charging loads the furnace charge into the furnace once or several times. Put in special containers, hoppers or baskets in advance. There are mainly two types of chain bottom plate type and clam type.

炉顶装料将炉料一次或分几次装入炉内。事先装入专门的容器，料斗或料筐。主要有链条底板式和蛤式两种类型。

Chain base plate type: the upper part is cylindrical, and the lower part is a row of triangular chain plates. The lower part is connected in series with chain or steel wire rope, and locked with a locking mechanism, forming a tank bottom. It is hung on the main hook of the crane and the buckle lock is hung on the auxiliary hook.

链条底板式：上部为圆筒形，下端是一排三角形的链条板，下端用链条或钢丝绳串联成一体，用扣锁机构锁住，并成一个罐底。吊在起重机主钩上，扣锁吊在副钩上。

Advantages: the charging tank can enter the furnace during charging, and it can be hoisted to the position 300mm from the furnace bottom to reduce the mechanical impact when the furnace charge falls.

优点：装料时料罐可进入炉膛内，吊至距炉底300mm的位置减轻了炉料下落时的机械冲击。

Disadvantages: the chain plates should be strung together again every time, with a large amount of maintenance.

缺点：链条板每次要重新串在一起，维护量大。

Clam type: grab type material tank, which can be divided into two open palates on both sides, two palates are closed by self weight, and the palates can be opened by the auxiliary hook of the crane through the lever system. It can control the falling speed of furnace charge, depending on the opening degree. It should be noted that the basket cannot be loaded into the furnace.

蛤式：抓斗式料罐，可分成两面向两侧打开的腭，两个腭靠自重闭合，用起重机的副钩通过杠杆系统可使腭打开。可靠打开程度控制炉料下落速度。料篮不能装入炉膛。

2.1.7 Auxiliary Equipment for Electric Furnace
2.1.7 电炉的辅助设备

2.1.7.1 Tilting Platform and Maintenance Platform
2.1.7.1 倾动平台与维修平台

Tilting platform: the tilting platform is a light-weight steel structure that is fixed on the cradle and can move with the cradle. When the furnace is in the smelting position, the tilting platform can fill the gap on the fixed platform.

倾动平台：倾动平台是一种轻型钢结构，固定在摇架上且能随摇架一起动作。在炉子位于冶炼位置时，倾动平台能填补固定平台上的空缺。

EBT bottom outlet (EBT) maintenance platform: EBT maintenance platform is also called outlet maintenance platform. It is telescopic for maintenance of steel outlets.

偏心底出钢口（EBT）维修平台：EBT 维修平台又称出钢口维修平台。其供维修出钢口用的是伸缩式的。

2.1.7.2 Furnace Wall Oxygen and Carbon Gun System
2.1.7.2 炉壁氧、碳枪系统

Taking a steel making plant as an example, there are 4 KT oxygen lances and 3 KT carbon lances in the upper and lower furnace shells of the electric arc furnace. Among them, there are two oxygen lances and one carbon gun in the upper furnace shell, and 1# oxygen gun and 1# carbon gun are installed on the 4# water cooling plate, and 4# oxygen gun is installed on the 14# water cooling plate; Two oxygen guns and two carbon guns are installed near the EBT area, below the water cooling plate of the upper furnace shell, 2# oxygen gun and 2# carbon gun are installed under the 6# water cooling plate, and 10# water cooling plate is installed 3# oxygen gun and 3# carbon gun.

以某炼钢厂为例，在电弧炉的上、下炉壳共有 4 支 KT 氧枪和 3 支 KT 碳枪。其中上炉壳有两支氧枪、一支碳枪，分别在 4 号水冷板上安装了 1 号氧枪和 1 号碳枪，14 号水冷板上安装了 4 号氧枪；下炉壳有两支氧枪、两支碳枪，安装在 EBT 区域附近，上炉壳水冷板的下方，6 号水冷板下方安装的是 2 号氧枪和 2 号碳枪，10 号水冷板下方安装的是 3 号氧枪和 3 号碳枪。

The arrangement of oxygen and carbon guns mainly considers raising the temperature in the cold zone of the furnace and the formation of foam slag. The purpose is to ensure a high oxygen penetration ability, accelerate the decarbonization reaction, and accelerate the melting of the steel material in the cold zone. The installation is shown in Figure 2-6.

氧、碳枪的布置主要考虑提高炉内冷区温度和泡沫渣的形成，目的是保证高的氧气穿透能力，加速脱碳反应，加速冷区钢铁料的熔化。安装如图 2-6 所示。

The supply of oxygen lance cooling water will be supplied by a dedicated atomizing water pump station. This system has a heat exchanger to exchange heat with the cooling water. Each gun

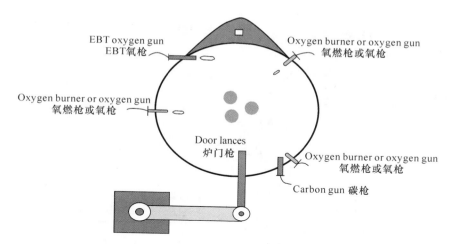

Figure 2-6　Arrangement of oxygen and carbon guns
图 2-6　氧、碳枪的布置

has an independent compressed air line. Compressed air and water are mixed at the entrance of the gun body and collected in the water tank of the atomizing water pump station. The separated gas is discharged from the exhaust of the water tank.

氧枪冷却水的供给将由专用雾化水泵站供给，这个系统带有热交换器，对冷却水进行热交换。每个枪有独立的压缩空气管线，压缩空气和水在枪体入口混合，收集在雾化水泵站的水箱内，分离出的气体从水箱的排气口排出。

The KT oxygen lance can be installed in a special cooling block. The gun body is always in a fixed position in the cooling block, no adjustment is required, and the same effect is ensured in the entire furnace body. Oxygen lances generate forces in the radial and tangential directions, causing natural agitation in the molten pool. The design speed of the KT oxygen lance is at least 2.1Mach, and the jet length is guaranteed to reach 1.7m.

KT 氧枪能安装在特殊的冷却块内，在冷却块内枪体总是在一个固定位置，不需要调节，在整个炉体中确保一样的效果。氧枪在径向和切线方向上产生力，使熔池产生自然的搅拌。KT 氧枪设计速度至少为 2.1Mach，喷射长度保证达到 1.7m。

The installation forms of KT oxygen and carbon guns are shown in Figure 2-7 ~ Figure 2-9.
KT 氧、碳枪安装形式如图 2-7 ~ 图 2-9 所示。

2.1.7.3　Furnace Door Oxygen Lance Equipment
2.1.7.3　炉门氧枪设备

The oxygen lance of the furnace door is centered on the column, one end of the cross arm is fixed on the rotating mast, three feed boxes are suspended on the other end of the cross arm, the feed box on the side of the column is the oxygen lance, and the other side is also for the oxygen gun, the middle is the carbon gun, as shown in Figure 2-10.

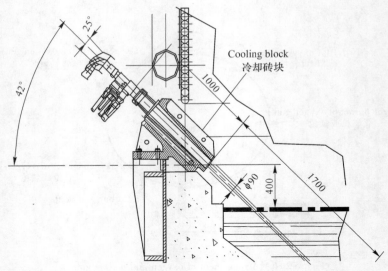

Figure 2-7　KT oxygen and carbon gun installation diagram
图 2-7　KT 氧、碳枪安装示意图

Figure 2-8　KT oxygen and carbon gun
图 2-8　KT 氧、碳枪

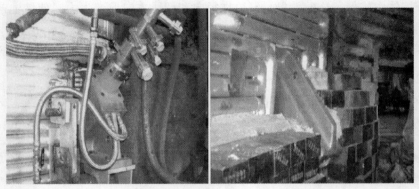

Figure 2-9　KT oxygen and carbon gun
图 2-9　KT 氧、碳枪

炉门氧枪以立柱为重心，横臂的一端固定在旋转立柱上，3个进给箱体悬挂在横臂的另一端，靠立柱侧的进给箱体为氧枪，另一侧的也为氧枪，中间的是碳枪，如图2-10所示。

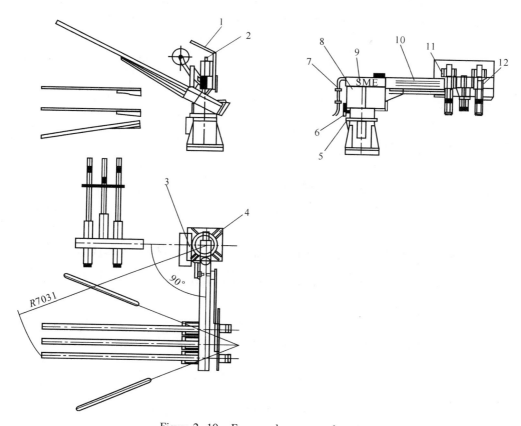

Figure 2-10　Furnace door oxygen lance

1—flameproof plate; 2—water cooling pipe; 3—horizontal swing mechanism;
4—column cross arm lifting and rotating mechanism; 5—cross arm rotating lifting limit;
6—lubricating oil centralized refueling device; 7—oxygen pipeline; 8—gas-liquid control box;
9—column housing; 10—gas-liquid pipeline; 11—oxygen feed mechanism; 12—carbon powder feed mechanism

图2-10　炉门氧枪

1—挡火板；2—水冷却管道；3—水平摆动机构；4—立柱横臂升降及旋转机构；
5—横臂旋转升降限位；6—润滑油集中加油装置；7—氧气管道；8—气-液控制箱；
9—立柱罩壳；10—气-液管道；11—氧气送进机构；12—碳粉送进机构

The action control of the furnace door oxygen lance is completed by both pneumatic and hydraulic parts. Among them, the feeding of the oxygen tube and the carbon powder tube and the rotation of the cross arm are completed by the pneumatic system, and the lifting and lowering of the cross arm, and the up and down and left and right swing of the feed box are controlled by hydraulic pressure.

炉门氧枪的动作控制由气动和液压两部分共同完成。其中氧气管和碳粉管的进给、横臂的旋转由气动系统完成，横臂的升降、进给箱的上下及左右摆动由液压控制。

Exercises

(1) What are the mechanical equipment of electric arc furnace?
(2) What are the requirements of the electric arc furnace for the furnace shell?
(3) What is the role of the reinforcement ring of the electric arc furnace shell?
(4) What are the advanced features of eccentric bottom tapping?
(5) What are the requirements of the electric arc furnace for the electrode lifting mechanism?

思考题

(1) 电弧炉有哪些机械设备？
(2) 电弧炉对炉壳有哪些要求？
(3) 电弧炉炉壳的加固圈有什么作用？
(4) 偏心底出钢的先进性表现在哪些方面？
(5) 电弧炉对电极升降机构有哪些要求？

Task 2.2　Electrical Equipment Preparation
任务2.2　电气设备准备

Mission objectives
任务目标

(1) Master the composition of electrical equipment.
(1) 掌握电气设备的构成。
(2) Master the requirements of the production process for electrical equipment.
(2) 掌握生产工艺对电气设备的要求。

2.2.1　Basic Function and Principle of Electric Equipment
2.2.1　电弧炉的电气设备的基本作用及原理

2.2.1.1　Basic Task
2.2.1.1　基本任务

The electric arc furnace uses three-phase alternating current as the power supply, works according to the principle of direct heating, uses the arc light to melt and smelt metals; the electric arc is produced by the strong current between the three graphite electrodes on the electric arc furnace and the furnace charge. EAF steelmaking is a process in which electric energy is converted into heat energy, which makes the charge melt for physical and chemical reaction. The main equipment to complete this energy conversion is the electric equipment of EAF. The main tasks of electric equipment of electric arc furnace are as follows:

(1) In order to obtain the strong current required by the process, EAF transformer is used to change the high voltage into the low voltage to provide electric energy.

(2) In order to keep the arc current and voltage current at a certain level, the arc furnace is equipped with an electrode automatic regulator, which can adjust the arc distance between the electrode and the charge to a certain value at any time during smelting. Therefore, EAF transformer and automatic regulating device of electrode lifting are the main electrical equipment of EAF.

电弧炉以三相交流电作为供电电源，按直接加热的原理工作，利用电弧的弧光熔化和冶炼金属；电弧是电弧炉上3个石墨电极与炉膛内炉料之间通过强大的电流而产生的。电弧炉炼钢是电能转变为热能而使炉料熔化进行物理化学反应的过程，而完成这种能量转变的主要设备就是电弧炉电气设备。电弧炉电气设备完成的任务主要有：

（1）为获得工艺要求的强大电流，用电炉变压器把高压变为低压，提供电能。

（2）为使电弧电流和电压电流保持在一定的水平上，电弧炉装有电极自动调节器，冶炼中随时调节电极和炉料之间的电弧距离为一定值。因此，电弧炉电气设备主要为电炉变压器和电极升降自动调节装置。

2.2.1.2　Working Principle
2.2.1.2　工作原理

The main circuit of EAF refers to the circuit from EAF transformer to EAF arc. It transforms high voltage into low voltage and high current to the electric arc furnace, provides the electric energy needed for smelting, and transforms the electric energy into heat energy in the form of electric arc.

电弧炉主电路指从电炉变压器供电到电炉电弧为止的电路。它将高电压转变为低电压大电流输给电弧炉，提供冶炼所需电能，并以电弧形式将电能转变为热能。

The working characteristic of EAF equipment is to turn on and off EAF transformer frequently, and the latter is done by high voltage circuit breaker. It can be seen that the high-voltage circuit breaker works under extremely heavy conditions. According to statistics, for the on-off frequency of EAF with multi-stage voltage, the on-off frequency in one year can reach 25000~30000 times.

炼钢电弧炉设备的工作特点是频繁接通和断开电炉变压器，后者的接通和断开通过高压断路器来完成。可见，高压断路器是在极其繁重的条件下工作的。据统计，对具有多级电压的无励磁电动调压变压器的电炉通断频度而言，一年期间通断次数共达2.5万~3万多次。

2.2.1.3　Electrical Characteristics
2.2.1.3　电气特性

The electric arc furnace uses UHV power distribution, which is sent to the furnace transformer through necessary high-voltage transformer, where it becomes the secondary voltage suitable for the occurrence of electric arc. The three-phase secondary circuit of the electric arc furnace is composed of the following equipment: furnace transformer; secondary conductor in the electrical room (also known as triangle sealing); flexible cable (mainly water-cooled cable); secondary conductor on the furnace (most recently the conducting arm); graphite electrode; alternating current arc (variable pure resistance).

电弧炉以特高压配电,经过必要的高压变压器送至炉用变压器,在这里变为适合于发生电弧的二次电压。电弧炉的三相二次电路由以下设备构成:炉用变压器;电气室内二次导体(也可称三角形封接);挠性电缆(主要为水冷电缆);炉上二次导体(最近以导电臂为主);石墨电极;交流电弧(可变纯电阻)。

The arc is a variable pure resistance, which is composed of resistance or inductive reactance. They have the function of arc stabilizers and contribute to the stability of the arc. The circuit of three-phase electric arc furnace is essentially a three-phase unbalanced circuit in terms of the composition and configuration of conductors, and the midpoint position of the three-phase circuit formed by scrap or molten steel is also not fixed, often carrying out complex migration. In addition, according to the characteristics of the arc and the heated scrap, the voltage and current change from sine wave to quite distorted waveform. Moreover, because of the continuous fluctuation of this state, the analysis of arc characteristics has to use statistical methods. Schematic diagram of electric arc furnace power supply circuit is shown as Figure 2-11.

电弧是可变纯电阻,而除此之外全部由电阻或感抗组成。它们具有电弧平稳器的机能,有助于电弧的稳定。三相电弧炉的电路,从导体的构成、配置看,本质上就是三相不平衡电路,并且以废钢或钢液形成的三相电路的中点位置也不固定,时常进行复杂的迁移。另外,根据电弧及被加热废钢的特性,电压及电流由正弦波变为相当畸变的波形。而且因为这个状态不断波动,电弧特性的解析不得不用统计方法。电弧炉供电线路原理如图 2-11 所示。

2.2.1.4 Power Supply Curve
2.2.1.4 供电曲线

The general goal of making the power supply curve of AC electric arc furnace is to smelt qualified molten steel with fast rhythm and low cost. The power supply curve formulated shall be able to operate safely and stably, taking into account the production rhythm at the same time, that is, to ensure that the apparent power borne by EAF transformer is not overloaded; the arc is stable and efficient burning ($0.75 \leq \cos\varphi \leq 0.86$); and the number of voltage on load switching shall be minimized.

制定交流电弧炉供电曲线的总目标是快节奏、低成本地冶炼出合格钢水。制定的供电曲线要能够安全、稳定运行,同时兼顾生产节奏,即:保证电炉变压器承受的视在功率不过载;电弧稳定高效燃烧($0.75 \leq \cos\varphi \leq 0.86$);电压有载切换次数尽可能少。

When making the power supply curve, the smelting characteristics and actual conditions shall be considered, and different power supply curves shall be made according to different raw material

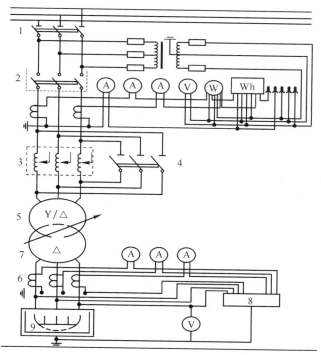

Figure 2-11 Schematic diagram of electric arc furnace power supply circuit
1—incoming disconnector; 2—high voltage circuit breaker; 3—external reactor;
4—reactor tap changer; 5—electric furnace transformer; 6—current transformer;
7—short net; 8—regulator; 9—electric arc furnace

图 2-11 电弧炉供电线路原理图
1—进线隔离开关；2—高压断路器；3—外加电抗器；4—电抗器分接开关；
5—电炉变压器；6—电流互感器；7—短网；8—调节器；9—电弧炉

structures and production requirements, power supply curve of 100t/60MV·A electric arc furnace for a certain raw material condition is shown in Figure 2-12. Generally speaking, the formulation of power supply curve is mainly considered from two aspects:

(1) Energy demand. To ensure the energy necessary for the melting and heating of the metal in the EAF at different stages.

(2) Effective use of energy. According to the characteristics of different stages of smelting, the favorable heating conditions and reasonable current, voltage and power are determined.

制定供电曲线时要考虑冶炼特点和实际条件，根据不同的原料结构和生产要求制定不同的供电曲线，100t/60MV·A 电弧炉某种原料状况用的供电曲线如图 2-12 所示。一般来说，制定供电曲线主要从两方面来考虑：

（1）能量需求。保证电弧炉冶炼过程中炉内金属在不同阶段熔化、升温时所必需的能量。

（2）能量的有效利用。针对冶炼不同阶段特点把握有利的加热条件、选定合理的电流、电压和功率。

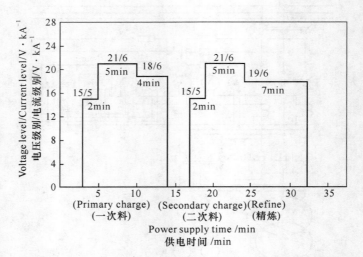

Figure 2-12 Power supply curve of 100t/60MV · A electric arc furnace for a certain raw material condition

图 2-12 100t/60MV·A 电弧炉某种原料状况用的供电曲线

2.2.2 Electric Equipment
2.2.2 电弧炉相关电气设备

2.2.2.1 Main Circuit of EAF
2.2.2.1 电弧炉的主电路

Large EAF steelmaking plants use a large amount of electricity, which is often required to be supplied by the transmission grid (110 ~ 220kV). In order to ensure the power supply, two lines or even two independent power supplies are often used to supply power to the bus of the substation in the steel plant. But in terms of power supply reliability, in addition to cooling water, EAF itself is allowed to cut off power suddenly, which will cause economic losses.

大型电炉炼钢厂用电量很大，常常要求由输电电网（110~220kV）供电。为了保证供电，常常由两条线路甚至两个独立电源向炼钢厂变电所的母线供电。但就供电可靠性讲，除了冷却水外，还是允许电炉本身突然停电的，当然这将引起经济损失。

The main circuit of EAF refers to the circuit from EAF transformer to EAF arc. It transforms high voltage into low voltage and high current to the electric arc furnace, provides the electric energy needed for smelting, and transforms the electric energy into heat energy in the form of electric arc.

电弧炉主电路指从电炉变压器供电到电炉电弧为止的电路。它将高电压转变为低电压大电流输给电弧炉，提供冶炼所需电能，并以电弧形式将电能转变为热能。

In order to cooperate with the normal progress of smelting, the electric arc furnace is also equipped with special mechanical equipment such as motor lifting device, tilting furnace mechanism, automatic adjustment and pressure change device. The electrical equipment serving these devices constitutes the auxiliary circuit of the furnace. The power of the auxiliary system is supplied by the substation of the whole plant.

为了配合冶炼的正常进行，电弧炉还设有电机升降装置、倾炉机构、自动调节和换压装置等专用的机械设备，服务于这些设备的电气设备，构成了炉子的辅助电路。辅助系统的电源由全厂的变电所供给。

The main circuit starts from the external power supply network and is in turn a disconnector, a high-voltage circuit breaker, a reactor, an EAF transformer, a short circuit network, and an arc.

主电路从外部供电网络开始依次为隔离开关、高压断路器、电抗器、EAF 变压器、短网、电弧。

2.2.2.2　Isolating Switch
2.2.2.2　隔离开关

Isolation switch is also called household switch, air short circuit switch. Isolation switches are mainly used to disconnect high-voltage power supplies during the maintenance of electric furnace equipment, and are sometimes used to switch operations.

隔离开关也称进户开关、空气短路开关。隔离开关主要用于电炉设备检修时断开高压电源，有时也用来切换操作。

Commonly used isolating switches are three-phase knife gate switches. The basic structure consists of insulators, knife gates, tie rods, shafts, handles and static contacts. This kind of switch has no arc extinguishing device, and it must be connected or cut off when there is no load. Therefore, the isolation switch must be operated after the high-voltage circuit breaker is opened. When the electric arc furnace is stopped or powered, the switch operation sequence is as follows: when the power is supplied, the isolation switch is closed first, and then the high voltage circuit breaker is closed; when the power is cut off, the high voltage circuit breaker is first opened, and then the isolation switch is opened. Otherwise, an arc may occur between the knife gate and the contacts, which may burn the equipment and cause a short circuit accident. In order to prevent misoperation, interlocking devices are often provided between the isolating switch and the high-voltage circuit breaker. The isolating switch cannot be operated when the high-voltage circuit device is closed. The operating mechanism of the isolating switch has three types: manual, electric and pneumatic. When performing manual operations, wear insulated gloves and stand on rubber pads to ensure safety.

常用的隔离开关是三相刀闸开关，基本结构由绝缘子、刀闸、拉杆、转轴、手柄和静触头组成。这种开关没有灭弧装置，必须在无负载时才可接通或切断电路，因此隔离开关必须在高压断路器断开后才能操作。电弧炉停止或送电时，开关操作顺序是：送电时先合上隔离开关，后合上高压断路器；停电时先断开高压断路器，后断开隔离开关。否则刀闸和触头之间产生电弧，会烧坏设备和引起短路事故等。为了防止误操作，常在隔离开关与高压断路器之间设有连锁装置，使高压电路器闭合时隔离开关无法操作。隔离开关的操作机构有手动、电动和气动三种。当进行手动操作时，应戴好绝缘手套，并站在橡皮垫上以保证安全。

2.2.2.3　High Voltage Circuit Breaker
2.2.2.3　高压断路器

　　The high-voltage circuit breaker is to make the high-voltage circuit turn on or off under load, and as a protection switch to automatically cut off the high-voltage circuit when the electrical equipment fails. High voltage circuit breakers used in electric arc furnaces include oil switches, air circuit breakers and vacuum circuit breakers.

　　高压断路器是使高压电路在负载下接通或断开，并作为保护开关在电气设备发生故障时自动切断高压电路。电弧炉使用的高压断路器有油开关、空气断路器和真空断路器。

　　(1) High-pressure oil switch. Oil switch is a kind of most common circuit breaker used to cut off or connect high voltage circuit. The contacts of the oil switch are immersed in transformer oil with good insulation performance, and the action of electric shock is completed by electromagnetic force. When the high-voltage circuit under load is disconnected, an arc will be generated between the contacts of the switching device, and the arc will destroy the contact point. The transformer oil decomposes and evaporates under the action of the arc to generate a large amount of steam, thereby cooling the arc and eliminating ionization to force the arc to extinguish. The oil also prevents contact oxidation, accelerates contact heat dissipation, and ensures insulation between the contact and the contact and ground.

　　(1) 高压油开关。油开关是用来切断或接通高压电路的一种最普通的断路器。油开关的触头浸泡在绝缘性能良好的变压器油中，触电的动作靠电磁力完成。当负载下的高压电路断开时，开关装置的各触点之间便会产生电弧，电弧对接触点有破坏作用。变压器油在电弧的作用下分解和蒸发而产生大量蒸汽，从而冷却电弧并消除电离迫使电弧熄灭。油还有防止触头氧化、加快触头散热、保证触头与触头之间以及对地的绝缘等作用。

　　The advantage of the oil switch is that it is easy to manufacture and cheap. However, an arc is generated for each action of the contact, which decomposes and carbonizes part of the oil, which reduces the insulating properties of the oil. At the same time, a large amount of gas is generated due to the decomposition of oil, 70% of which is explosive hydrogen. The oil switch contacts need to be changed and repaired every 1000 times.

　　油开关的优点是制造方便，价格便宜。但由于触点每动作一次都要产生电弧，使部分油料分解碳化，致使油的绝缘性能降低。同时油的分解还会产生大量气体，其中70%是易爆燃的氢气。油开关触头每动作1000次左右需进行换油和检修触头。

　　(2) Electromagnetic air circuit breaker. Electromagnetic air circuit breaker, also known as magnetic blowing switch, its arc extinguishing device is magnetic blowing solenoid type. The three-phase high-voltage power supply is respectively connected to the tail of the static contact and the moving arc outlet, and installed side by side on the same foundation. When the contacts are sepa-

rated to generate an arc, the gas ejected from the compressed air bladder and the double π shaped magnetic system will cause the arc to rise rapidly, elongate into the arc solenoid, and be cooled by the arc isolation plate and atmospheric medium and finally extinguished.

（2）电磁式空气断路器。电磁式空气断路器，又称磁吹开关，其灭弧装置为磁吹螺管式。三相高压电源分别连接静触头和动弧触头的尾部，且并列安装在同一基础上。当触头分开产生电弧后，便在压气皮囊喷出的气体和双π形磁系统的作用下，使电弧迅速上升、拉长进入电弧螺管，并受到隔弧板和大气介质的冷却最后熄灭。

Compared with the oil switch, the air circuit breaker has a short switching time, does not cause fire, explosion and overvoltage, and it is stable and reliable, and is suitable for frequent operation.

空气断路器与油开关相比，其开关时间短，不会发生起火、爆炸及产生过电压现象，工作平稳可靠，适于频繁操作。

（3）Vacuum circuit breaker. Vacuum circuit breaker is a relatively advanced high-voltage circuit switch that can better meet the requirements of increasing power. The main part of this switch is a vacuum chamber, the contacts are made of high melting point metals W, Mo, Ti, etc., because the arc voltage generated in the vacuum chamber is very low, and the charged particles in the arc are easy to extend outside the arc diffusion, little energy consumption at the contacts, reduces the evaporation of metal contacts, and extends the life of the switch. It can be safely used 40000 times on the arc, and its life is 40 times longer than that of the oil switch.

（3）真空断路器。真空断路器是一种比较先进的高压电路开关，可以较好地满足功率不断增大的要求。这种开关的主要部件是一个真空室，触头由高熔点金属W、Mo、Ti等做成，由于在真空室内产生的电弧电压很低，且电弧中的带电质点在真空下易于向弧体外扩散，触点处能量消耗很少，减少了金属触点的蒸发量，延长了开关寿命。它在电弧上可安全使用40000次，寿命比油开关长40倍。

Vacuum circuit breakers should always keep the insulation parts clean. When the vacuum interrupter opens, the arc light changes from blue to red. The vacuum interrupter should be replaced in time.

真空断路器应经常保持绝缘件的清洁，当真空灭弧室分闸时的弧光由蓝色变为红色，应及时更换真空灭弧室。

2.2.2.4 Reactor
2.2.2.4 电抗器

The reactor is connected in series on the high voltage side of the transformer. Its function is to increase the inductive reactance in the circuit to achieve the purpose of stabilizing the arc and limiting the short circuit current.

电抗器串联在变压器的高压侧，其作用是使电路中感抗增加，以达到稳定电弧和限制短路电流的目的。

During the melting period of EAF, the electrode is often short circuited due to the collapse of the material. The short-circuit current is often many times higher than the rated current of the transformer, which leads to the decrease of the service life of the transformer. After connecting the reactor, when the current value in the main circuit increases, an induced current will be generated in the reactor winding. The magnetic field generated by the induced current will hinder the increase of the main circuit current, so it can effectively limit the short-circuit current. On the contrary, when the current in the main circuit decreases, the magnetic field produced by the reactor winding will hinder the decrease of the current in the main circuit. Generally, the short-circuit current after connecting the reactor is not more than 2.5~3.5 times of the rated current. In this current range, the automatic adjustment device of the electrode can ensure that the lifting motor reduces the load without tripping and power failure, and also plays a role of stabilizing the arc.

在电弧炉炼钢的熔化期经常由于塌料而引起电极的短路，短路电流常超过变压器额定电流的许多倍，导致变压器寿命降低。接入电抗器后，当主回路中电流值增大时，电抗器绕组中便产生一感应电流，由感应电流产生的磁场将阻碍主回路电流的增大，故能有效地限制短路电流。反之，当主回路中的电流值减小时，电抗器绕组产生的磁场将阻碍主回路电流的减小。一般在接入电抗器后的短路电流都不超过 2.5~3.5 倍的额定电流，在这个电流范围内，电极的自动调节装置能够保证提升电极降低负载，而不致跳闸停电，也起到了稳定电弧的作用。

The reactor of the small furnace can be installed inside the transformer box of the electric furnace. Large furnaces are equipped with reactors separately. For the electric furnace above 20t, the reactance percentage of the main circuit itself is already very large, so no special reactor is needed.

小炉子的电抗器可装在电炉变压器箱体内部。大炉子则单独设置电抗器。对 20t 以上的电炉子则因主电路本身的电抗百分数已经很大，无须专门设置电抗器。

2.2.2.5　EAF Transformer
2.2.2.5　电弧炉变压器

EAF has the characteristics of low arc voltage and high current, which requires a special EAF transformer to connect with the high voltage grid. Other electrical equipment is centered on EAF transformer and serves the latter. The EAF transformer is characterized by a very large current on the low voltage side, so it should be installed near the EAF supplied by it.

电炉电弧电压低而电流大的特性，要求有专用的电炉变压器和高压电网相连接。其他电气设备是以电炉变压器为中心，并为后者服务的。电炉变压器的特征是低压侧电流非常大，因此应将它安装在由它供电的电炉附近。

Features of EAF transformer: generally, the original side of EAF transformer is powered by 6~35kV power supply. But at present, there are EAF transformers which are directly supplied by the transmission network of more than 60kV or even 220kV. The low voltage and high current outgoing line of EAF transformer is an important part of it. The outgoing line shall be prevented from over heating of itself and adjacent structures. The outgoing line shall be protected from oil leakage and bear the mechanical force in case of short circuit. In addition, the impedance between the three-phase outgoing lines of the transformer is required to be the same as much as possible.

电炉变压器的特点：通常电炉变压器原边由 6~35kV 电源供电。但目前已有由 60kV 以上甚至 220kV 的输电电网直接供电的电炉变压器。电炉变压器的低压大电流出线是其极为重要的一部分。出线要防止本身和邻近结构的过分发热。出线处要防止漏油并要承受短路时的机械力量。此外，要求变压器三相出线间的阻抗尽可能相同。

2.2.2.6 Short Network
2.2.2.6 短网

From the outgoing line at the voltage side of EAF transformer to the arc in the furnace, conducting low voltage and high current conductor is called short network. In terms of electrical properties, graphite electrodes are also included. On the steel-making electric arc furnace, because of the impact of short-circuit current, there will be a lot of electric power between the conductors. Therefore, only copper with better mechanical strength is used as conductor. The size of conductor section depends on the current. For naturally cooled conductors, the current density shall not be greater than $1.5 A/mm^2$. The current density of water-cooled conductor can be $3.5 \sim 5.0 A/mm^2$. For the current of each phase up to tens of kiloamperes, the water-cooled cable with a cross-section of $3000 \sim 6000 mm^2$ is used, and the allowable current density is up to $10 A/mm^2$. Only two water-cooled cables are used for each phase. The resistance and reactance of the short network have a great influence on the electrical characteristics of EAF. People always try to reduce the value of resistance and reactance, and try to make the values of three phases equal.

从电炉变压器压侧的出线开始到炉中电弧为止，传导低电压、大电流的导体，称为短网。从电气特性看来，石墨极也包括在内。炼钢电弧炉上，由于经常发生冲击性的短路电流，导体间会有很大的电动力，因而只采用机械强度较好的铜作为导体。导体截面大小取决于电流大小，对自然冷却导体，电流密度不大于 $1.5 A/mm^2$；对水冷导体电流密度可取 $3.5 \sim 5.0 A/mm^2$。对每相电流大至几十千安，采用每根截面达 $3000 \sim 6000 mm^2$ 的水冷电缆，允许电流密度高达 $10 A/mm^2$，每相只用两根水冷电缆。短网的电阻和电抗对电炉电气特性有重大影响。人们总是力图减小其电阻和电抗的数值，并且力图使三相的数值相等。

The structure of the short network is mainly composed of three parts: hard copper bus bar (copper bar), flexible cable and top water-cooled copper tube.

短网的结构主要由硬铜母线（铜排）、软电缆和顶水冷铜管三部分组成。

Use a piece of flexible cable to connect the output end of the secondary winding of the transformer to the hard copper bar (about 400mm long, the wire diameter is the same as the flexible cable after the hard copper bar). The advantages are:

(1) Reduce the vibration of the connection with the transformer and cause the connection screw to loosen, reduce the resistance, and increase the output power;

(2) Reduce the leakage of the transformer due to vibration.

从变压器副边绕组出线端与硬铜排之间采用一段软电缆连接（长约400mm、线径与硬铜排之后的软电缆相同)。其优点是：

(1) 减小与变压器连接处的震动而造成连接螺丝松脱，减小电阻，提高输出功率；

(2) 减小变压器因震动而漏油。

Flexible cable is also called soft bus, and its head and tail are connected with copper bar and water-cooled conductive copper tube. The length of the flexible cable is limited to meet the electrode lifting and furnace tilting. The flexible cable is made of thin copper wire per share, and strives to have a larger surface area. According to the rated current of the transformer, multiple flexible cables are connected in parallel. The flexible cable is generally bare copper cable, and the allowable current density is $1.0 \sim 1.5 A/mm^2$. If the bare copper cable is covered with water-cooled rubber tabe, the allowable current density can be increased by $2 \sim 3$ times, at the same time, it can save the number of cables and improve the service life.

软电缆又称软母线，其首尾与铜排及水冷导电铜管相连。软电缆的长度以能满足电极升降和炉体倾动为限。软电缆由每股细铜线绕成，力求有较大的表面积，根据变压器额定电流的大小，采用多根软电缆并联连接。软电缆一般为裸铜电缆，允许的电流密度为 $1.0 \sim 1.5 A/mm^2$，如在裸铜电缆外套水冷胶管，允许的电流密度可以提高 $2 \sim 3$ 倍，同时可起到节约电缆根数、提高使用权寿命的效果。

In recent years, the newly developed large-section flexible water-cooled cable has been applied in the short network of the electric furnace. It is a circle of each phase of the water-cooled cable, which is fixed by a hose inside and outside and cooled by water. The advantages of this kind of cable are: it can restrain electromagnetic vibration, prevent wear and tear, increase service life by many times; it can reduce impedance, operation stably, allow current density of $4.5 A/mm^2$, increase transformer output by $10\% \sim 15\%$, save electricity by $3\% \sim 5\%$, and reduce furnace lining material buming loss.

近年来新研制的大截面柔性水冷电缆已在电炉短网上应用，它是将一相的各股水冷电缆组成圆形，内外由胶管固定并通水冷却。这种大截面集束电缆的优点是：能阻抑电磁振动，防止磨损，使用寿命成倍提高；阻抗减少，运行稳定，允许电流密度为 $4.5 A/mm^2$，使变压器出力提高 $10\% \sim 15\%$，节电 $3\% \sim 5\%$，并可降低炉衬材料烧损。

The water-cooled conductive copper tube is installed above the electrode cross arm, and is connected to the flexible cable and the electrode clamp head from end to end. Each phase electrode has two water-cooled conductive copper tubes. The thickness of the tube arm is generally $10 \sim 15mm$. The tube is cooled by water, and the allowable current density is $3.5 \sim 6.0 A/mm^2$.

水冷导电铜管装在电极横臂的上方，首尾与软电缆及电极夹头相连。每相电极有两根水冷导电铜管，管臂厚度一般为 $10 \sim 15mm$，管内通水冷却，允许的电流密度为 $3.5 \sim 6.0 A/mm^2$。

2.2.2.7　Electrode

2.2.2.7　电极

The name of electrode comes from the fact that it is the extreme of main circuit of EAF. In the steelmaking arc furnace, the electrode directly contacts the high power arc, so the working condition of the electrode is bad. High temperature gradient and thermal stress are produced in the longitudinal and transverse direction of the electrode. When the charge collapses during melting, it may be impacted by the block. In production, carbon material can only be used as electrode in EAF. Artificial graphite electrode is often used. Graphite has the following properties:

(1) After heating, graphite sublimates directly from solid state to gas state, and the sublimation temperature is as high as 3800℃.

(2) It is different from most materials. When the temperature rises, its mechanical strength increases.

(3) Compared with other materials, graphite has good thermal conductivity and low expansion coefficient, which makes the thermal shock resistance of graphite better and reduces the thermal stress in the electrode.

(4) When the surface temperature of graphite is higher than 400℃, it will combine with oxygen. The amount of oxidation is related to the oxygen content, gas flow rate and exposure time. When the temperature is higher than 600℃, the oxidation process will become very intense.

(5) Graphite is easy to be machined, which can make the screw joint seat and screw joint at both ends of the electrode have high machining accuracy, good contact at the electrode joint and good mechanical stress distribution.

(6) Compared with some high melting point metals, such as tungsten, molybdenum and tantalum, graphite is much cheaper.

电极的名称来源于它是炼钢电弧炉主电路的极端。炼钢电弧炉中，电极的工作条件是恶劣的，电极直接接触大功率电弧。在其纵向和横向产生很高的温度梯度和热应力。在炉料熔化过程中塌料时，有可能遭到料块的机械撞击。生产中，炼钢电弧炉只能采用碳素材料作为电极，常采用人造石墨化电极。石墨具有下列性能：

(1) 石墨加热后，直接由固态升华为气态，升华温度高达3800℃。

(2) 它与大部分材料不同，在温度上升时，其机械强度上升。

(3) 相对于其他材料，石墨的导热性能好而膨胀系数较低，使石墨的抗热震性能较好，降低了电极中的热应力。

(4) 在石墨表面温度大于400℃后会和氧气相结合。氧化量与气体中的氧含量、气体流速和暴露时间有关。在温度大于600℃后，氧化过程将变得很激烈。

(5) 石墨易于机械加工，可使电极两端的螺纹接头座和螺纹接头有较高的加工精度，电极接头处接触良好，机械应力分布较好。

(6) 与某些高熔点金属（如钨、钼、钽等）相比，石墨的价格便宜很多。

Electrode consumption generally has the following three conditions:

(1) Electrode end consumption. Continuous sublimation under the high temperature of the arc; due to the extremely high temperature of the arc, the colder scrap in the melting process of the furnace charge will make the electrode cool, and the thermal stress generated in the electrode will make the electrode peel off; under the high current, the arc will shift outward sharply. When

the slag layer is thick, the electrode end contacts with the slag liquid phase and is partially melted.

(2) Electrode side consumption refers to the oxidation consumption of electrode cylinder surface. When the temperature is over 400℃, oxygen can penetrate into the surface of graphite and oxidize. When the temperature is over 550~600℃, the oxidation of the electrode surface is intensified. The oxygen concentration and gas velocity near the electrode have a great influence on the oxidation loss of the electrode. Generally, oxygen fuel burners, oxygen blowing operation, dust removal and exhaust will affect electrode consumption.

(3) Electrode breaking loss. In the process of charge melting, the loss of electrode breaking caused by the falling of charge tower. The manufacturing process of graphite electrode is complex and the production cycle is long. Therefore, the price of graphite electrode in EAF is higher, especially in the EAF with large diameter and ultra-high power. The electrode consumption directly affects the cost of EAF steelmaking. In EAF steelmaking, electrode consumption should be reduced as much as possible, so as to have low cost and improve the market competitiveness of enterprises.

电极消耗一般有下列3种情况：

(1) 电极端面消耗。在电弧高温下不断升华；由于电弧温度极高，在炉料熔化过程中较冷的废钢会使电极冷却，在电极中产生的热应力使电极剥落；在大电流下电弧剧烈向外偏移。在渣层较厚时，电极端和渣液相接触而被部分熔解。

(2) 电极侧面消耗是指电极圆柱体表面被氧化消耗。在温度超过400℃时，氧气即能渗入石墨表面发生氧化，在温度超过550~600℃时电极表面氧化加剧。电极附近的氧浓度和气流速度对电极氧化损失有较大影响。一般氧气燃料烧嘴、吹氧操作和除尘排气等，都会影响电极消耗。

(3) 电极折断损失。在炉料熔化过程中由于炉料塌落造成的电极折断损失。石墨电极制造过程复杂、生产周期长，因此，炼钢电弧炉的石墨电极价格较高，特别是直径大的超高功率电弧炉所使用的电极价格更高。电极消耗的指标直接影响电弧炉炼钢生产成本。电弧炉炼钢中，要尽可能降低电极消耗，这样才能有低的成本，提高企业的市场竞争力。

2.2.2.8　EAF Load
2.2.2.8　电弧炉负载

From the perspective of the importance of power supply, the steel-making electric arc furnace will not cause personal and major equipment accidents due to sudden shutdowns, and allows short-term power outages. From the perspective of electricity consumption, the magnitude of the load is very different at different stages of the steelmaking process. It is only during the melting period of the charge that the electric furnace transformer is fully loaded. During the preparation period for tapping and supplementary furnace charging, the furnace is powered off. The load is lighter during refining. The annual full-load utilization time of the electric furnace transformer is less than 2500 hours. If the electric furnace is only used as a tool for melting scrap steel, it can be increased to 3500 hours.

从供电重要程度来看,炼钢电弧炉不会由于突然停电而造成人身和重大设备事故,允许短时停电。从用电角度来看,在炼钢过程的不同工艺阶段,负载的大小是大不相同的。只是在炉料熔化期中,电炉变压器才是满载工作。在出钢、补炉装料的预备期间,炉子停电。精炼期间负载较轻。电炉变压器的全年满载利用时间小于2500h。若将电炉只作为熔化废钢的工具,则可提高至3500h。

The average power factor of EAF is usually only 0.70 ~ 0.85, and the value of large furnace is smaller. As we all know, the electric current of the electric arc furnace does not change in breaking ground. Especially in the melting stage of the charge, the position of the arc moves continuously with the melting of the charge. The arc can suddenly jump from one charge to another that becomes closer. The shape of the charge is very irregular, which makes the arc length constantly changing. Cold furnace material and slag material change the atmosphere around the arc, which is the source of arc current change. After the formation of the molten pool and during the refining period, the current is much more stable.

电炉的平均功率因数常常只有0.70 ~ 0.85,大炉子的值要小些。众所周知,电弧炉的电流在不断变动,尤其是在炉料的熔化阶段,电弧位置随某块炉料熔化而连续移动。电弧能从一块炉料突然跳至距离变得更近的另一块炉料。炉料形状是极不规则的,使电弧长度不断变动。冷炉料、渣料等又使电弧四周气氛改变,这些是电弧电流变动的根源。在金属熔池形成后以及在精炼期中,电流稳定得多。

Exercises

(1) Please analyze the role of electric arc furnace electrical equipment.

(2) What are the electrical equipment of electric arc furnace?

(3) What are the requirements of the electric arc furnace for transformer oil?

(4) What are the losses of the electrode?

思考题

(1) 请分析电弧炉电气设备的作用。

(2) 电弧炉电气设备都包括哪些?

(3) 电弧炉对变压器有哪些要求?

(4) 电极有哪几方面的损耗?

Project 3　Electric Furnace Steelmaking Production
项目3　电炉炼钢生产

Task 3.1　Operation of Charging and Batching
任务3.1　补炉配料装料操作

Mission objectives
任务目标

(1) Understand the material, method and principle of furnace mending.
(1) 了解补炉材料、补炉方法、补炉原则。
(2) Master the ingredient requirements.
(2) 掌握配料要求。
(3) Master the loading method.
(3) 掌握装料方法。

3.1.1　Make Up the Stove
3.1.1　补炉

3.1.1.1　'Three Factors' Affecting Lining Life
3.1.1.1　影响炉衬寿命的"三要素"

(1) Type, nature and quality of furnace lining.
(2) High temperature arc radiation and chemical erosion of slag.
(3) Oxygen blowing operation and slag, steel and other mechanical scour as well as charging impact.
(1) 炉衬的种类、性质和质量。
(2) 高温电弧辐射和熔渣的化学侵蚀。
(3) 吹氧操作与渣、钢等机械冲刷以及装料的冲击。

3.1.1.2　Furnace Repair Position
3.1.1.2　补炉部位

The slag line of furnace wall is seriously damaged by the radiation of high temperature arc, chemical erosion and mechanical erosion of slag and steel, and oxygen blowing operation.

The hot spot area of slag line, especially the No. 2 hot spot area, is seriously eroded by the influence of high arc power and arc deflection. The damage degree of this point often becomes the basis for furnace change.

Near the tap hole, it is easy to thin due to the erosion of slag steel.

Both sides of the furnace door are often damaged by the action of quench and heat, the scouring of slag and the collision between operation and tools.

炉壁渣线，受到高温电弧的辐射，渣、钢的化学侵蚀与机械冲刷，以及吹氧操作等损坏严重。

渣线热点区，尤其2号热点区还受到电弧功率大、偏弧等影响侵蚀严重，该点的损坏程度常常成为换炉的依据。

出钢口附近，因受渣钢的冲刷也极易减薄。

炉门两侧，常受急冷急热的作用、流渣的冲刷及操作与工具的碰撞等，损坏也比较严重。

3.1.1.3　Mending Method
3.1.1.3　补炉方法

The furnace mending methods are divided into manual mending and mechanical gunning. According to different mixing methods of selected materials, they are divided into dry mending and wet mending. The principle of furnace mending is: high temperature, fast mending and thin mending.

补炉方法分为人工投补和机械喷补，根据选用材料的混合方式不同，又分为干补和湿补两种。补炉的原则是高温、快补、薄补。

3.1.1.4　Furnace Mending Material
3.1.1.4　补炉材料

The mechanical gunning material is mainly composed of magnesia, dolomite or a mixture of the two, and is mixed with binder such as phosphate or silicate.

机械喷补材料主要用镁砂、白云石或两者的混合物，并掺入磷酸盐或硅酸盐等黏结剂。

3.1.2　Ingredients
3.1.2　配料

The primary task of batching is to ensure the smooth operation of smelting. Scientific batching should not only be accurate, but also use iron and steel materials reasonably. At the same time, it should ensure to shorten smelting time, save alloy materials and reduce consumption of metal and other auxiliary materials.

配料的首要任务是保证冶炼的顺利进行。科学的配料既要准确，又要合理地使用钢铁料，同时还要确保缩短冶炼时间、节约合金材料并降低金属及其他辅助材料的消耗。

3.1.2.1 Basic Requirements for Ingredients
3.1.2.1 对配料的基本要求

(1) Accurate ingredients. Generally, the ingredients are made according to the steel grades, equipment conditions, existing raw materials and different smelting methods. The accuracy of burden includes burden weight and burden composition. If the burden weight is not accurate, it will easily lead to improper control of chemical composition in the smelting process, or the ingot will be short of support and waste products, or excessive injection surplus may occur and increase consumption. If the chemical composition of the charge is not matched correctly, it will bring great difficulties to the smelting operation and even make the smelting impossible. Taking oxidation smelting as an example, if the carbon content is too high, the ore consumption will be increased or the oxygen consumption time will be prolonged; if the carbon content is too low, the carbon will be increased after melting; if the content of non oxidation elements is higher than the specification of smelting steel, it is necessary to add other metal materials to remove the excess content or carry out steel modification, which not only prolongs the smelting time, reduces the service life of furnace lining, but also increases the service life of each lining. The consumption of raw materials affects the quality of steel. If the steel is too high and there is no other steel grade to be changed, smelting can only be stopped. In order to eliminate the above situation, it is necessary to master the chemical composition of steel and ferroalloy before batching.

(1) 准确配料。一般是根据冶炼的钢种、设备条件、现有的原材料和不同的冶炼方法进行配料。配料的准确性包括炉料重量及配料成分两个方面。配料重量不准，容易导致冶炼过程化学成分控制不当或造成钢锭缺支短尺废品，也可能出现过量的注余增加消耗。炉料化学成分配得不准，会给冶炼操作带来极大的困难，严重时将使冶炼无法进行。以氧化法冶炼为例，如配碳量过高，会增加矿石用量或延长用氧时间；配碳量过低，熔清后势必进行增碳；配入不氧化元素的含量如果高于冶炼钢种的规格，需加入其他金属料撤掉多余的含量或进行改钢处理，既延长了冶炼时间，降低了炉衬的使用寿命，增加了各种原材料的消耗，又影响钢的质量，如果配得过高而又无其他钢种可更改时，只有终止冶炼。为了杜绝以上情况的发生，配料前掌握有关钢铁料及铁合金的化学成分是十分必要的。

In fact, there are many factors that affect the accuracy of batching, not only related to planning, calculation and measurement, but also related to the yield rate, furnace condition, scientific management of steel and ferroalloy, operation level of charging and steelmaking workers, etc.

实际上，影响配料准确性的因素较多，除与计划、计算及计量有关外，还与收得率、炉体情况、钢铁料及铁合金的科学管理、装料工和炼钢工的操作水平等有关。

(2) Use principle of steel material. The use principle of iron and steel material mainly includes smelting method, charging method, chemical composition of steel type and quality requirements of products. According to the different characteristics of smelting methods, the chemical composition of steel must meet the needs of smelting steel. Oxidation method has better ability of dephosphorization, degassing and inclusion removal, so common coarse materials should be used

more; back blowing method and non oxidation method can recover valuable alloy elements due to their weak ability of dephosphorization, degassing and inclusion removal, so high-quality returned fine materials should be used as much as possible. Due to the high quality and service performance requirements for bearing steel, crankshaft steel and high standard structural steel, no matter what method is adopted for smelting, it is better to use more refined materials.

（2）钢铁料的使用原则。钢铁料的使用原则主要应考虑冶炼方法、装料方法、钢种的化学成分以及产品对质量的要求等。根据冶炼方法的不同特点使用钢铁料，钢铁料的化学成分必须符合冶炼钢种的需要。氧化法有较好的脱磷、去气、除夹杂的能力，应多使用普通的粗料；返吹法和不氧化法因脱磷、去气、除夹杂能力不强，但能回收贵重的合金元素，所以应尽量使用优质的返回精料。由于对轴承钢、曲轴钢以及高标准的结构钢等的质量与使用性能要求较高，无论采用何种方法冶炼，最好多用一些精料。

In addition, the briquette and unit volume weight of iron and steel should be mastered in advance. In general, 30%~40% of the bulk material, 40%~50% of the medium bulk material and 15%~25% of the small or light iron shall be added to the furnace charge. Of course, when the material source is not good or the refining outside the furnace is used, light and thin mixed iron can also be mixed. During manual charging, the briquette and weight of the steel material must be suitable for the size and manpower of the furnace door, and the light and thin materials should not be too much, so as to avoid prolonging the charging time. During the mechanical charging of furnace top, due to the use of mechanical equipment and the full use of smelting room space, larger heavy materials and more light and thin materials can be used.

此外，在配料时，还应预先掌握钢铁料的块度和单位体积重量。一般炉料中应配入大块料30%~40%、中块料40%~50%、小块料或轻薄铁15%~25%。当然，料源不好或采用炉外精炼时，轻薄杂铁也可多配。人工装料时，钢铁料的块度及重量必须与炉门的尺寸和人力相适应，轻薄料也不宜过多，以免延长装料时间。炉顶机械装料时，由于采用机械设备且能充分利用熔炼室空间，可使用较大的重料及较多的轻薄料。

3.1.2.2 Proportioning Calculation Formula
3.1.2.2 配料计算公式

(1) Principle of charge composition. In the batching process, the chemical composition of furnace charge mainly considers the specification and composition of steel, smelting method, element characteristics and specific requirements of process, etc. Specifically:

（1）炉料成分的配定原则。配料过程中，炉料化学成分的配定主要考虑钢种规格成分、冶炼方法、元素特性及工艺的具体要求等。具体为：

1) Carbon allocation. The carbon allocation in the charge mainly considers the specification and composition of the steel, the burning loss of carbon in the melting period and the decarburization amount in the oxidation period, as well as the addition of alloy and slag making system in the reduction period. The burning loss of carbon element in melting period is related to the way of melting aid. According to the specific conditions of actual production, the inherent law can be

summarized, and the general fluctuation is about 0.60%. The decarburization amount in oxidation period shall be determined according to the specific requirements of the process. For the first furnace in new furnace, the decarburization amount shall be greater than 0.40%. The setting of non oxidizing carbon should ensure that the full melting carbon is near the lower limit of steel specification.

1) 碳的配定。炉料中碳的配定主要考虑钢种规格成分、熔化期碳的烧损及氧化期的脱碳量，还应考虑还原期补加合金和造渣制度对钢液的增碳。熔化期碳元素的烧损与助熔方式有关，可根据实际生产的具体条件，总结固有规律，一般波动在0.60%左右。氧化期的脱碳量应根据工艺的具体要求而定，对于新炉时的第一炉，脱碳量应大于0.40%。不氧化法碳的配定应保证全熔碳位于钢种规格要求的下限附近。

2) Silicon coordination. In general, the silicon of iron and steel smelting by oxidation method is mainly brought in by pig iron and scrap, and the silicon after full melting shall not be more than 0.30%, so as to avoid delaying the boiling time of molten pool. In order to improve the recovery rate of alloy elements, scrap silicon steel or ferrosilicon can be added according to the process requirements, but it should not be more than 1.0%, or not for special cases.

2) 硅的配定。在一般情况下，氧化法冶炼钢铁料的硅主要是由生铁和废钢带入，全熔后的硅不应大于0.30%，以免延缓熔池的沸腾时间。返吹法冶炼为了提高合金元素的收得率，根据工艺要求可配入硅废钢或硅铁，但也不宜超过1.0%以上，对于特殊情况也可不配。

3) Determination of manganese. For the steel grades smelted by oxidation method, if the specification content of manganese is high, it will not be taken into account when batching; if the specification content of manganese is low, it should be strictly controlled when batching, so as to avoid the operation of manganese removal by steelmakers as much as possible. For some important steel grades, in order to make the non-metallic inclusions in the steel fully float, the manganese content in molten steel after melting should not be less than 0.20%, but should not be too high, so as not to affect the pool boiling and dephosphorization. Since it is difficult to remove manganese by non oxidation or back blowing smelting, the manganese content shall not exceed the medium limit of steel specification. Manganese in high speed steel affects the grain size of steel, and the lower the content of manganese, the better.

3) 锰的配定。用氧化法冶炼的钢种，如锰的规格含量较高，配料时一般不予以考虑；如锰的规格含量较低，配料时应严格控制，尽量避免炼钢工进行脱锰操作。对于一些用途重要的钢种，为了使钢中的非金属夹杂物能够充分上浮，熔清后钢液中的锰含量不应低于0.20%，但也不宜过高，以免影响熔池的沸腾及脱磷。由于不氧化法或返吹法冶炼脱锰操作困难，因此配锰量不得超过钢种规格的中限。高速钢中锰影响钢的晶粒度，配入量应越低越好。

4) Chromium matching. The chromium content in the steel smelted by oxidation method should be as low as possible. When smelting high chromium steel, the non oxidation method of chromium distribution is controlled according to the middle and lower limit of steel output, and the

back blowing method is lower than the lower limit.

4）铬的配定。用氧化法冶炼的钢种，钢中的铬含量应尽可能地低。冶炼高铬钢时，配铬量不氧化法按出钢量的中下限控制，返吹法则低于下限。

5) The coordination of nickel and molybdenum. When the content of nickel and molybdenum in the steel is high, the content of nickel and molybdenum shall be mixed according to the lower and middle limits of the steel specification, and installed together with the furnace charge. When smelting non nickel steel, the nickel content in the steel material shall be lower than the residual composition specified in the steel grade. Nickel in high speed steel is harmful to hardness, so the lower the residual content is, the better.

5）镍、钼元素的配定。钢中镍、钼含量较高时，镍、钼含量按钢种规格的中下限配入，并同炉料一起装炉。冶炼无镍钢时，钢铁料中的镍含量应低于该钢种规定的残余成分。高速钢中的镍对硬度有害无利，因此要求残余含量越低越好。

6) Determination of tungsten. Tungsten is a weak reducing agent. In the process of steel smelting, there are different losses due to different ways of using oxygen. In ore smelting, no steel is artificially prepared with tungsten, and the lower the residual tungsten, the better. When smelting by non oxidation method and back blowing method, it shall be mixed according to the lower and middle limit of steel specification content, and installed together with furnace charge. Molybdenum in many tungsten steels can replace part of tungsten in composition, which should be paid more attention in the process of batching.

6）钨的配定。钨是弱还原剂，在钢的冶炼过程中，因用氧方式的不同而有不同的损失。矿石法冶炼，任何钢种均不人为配钨，且要求残余钨量越低越好。不氧化法和返吹法冶炼时，应按钢种规格含量的中下限配入，并同炉料一起装炉。许多钨钢中的钼在成分上可代替部分钨，配料过程中应严加注意。

7) The principle of the composition of the charge of the brush pot steel. In the electric furnace steelmaking workshop, after the smelting of high alloy steel containing Cr, Ni, Mo, W or Mn, etc., it is necessary to smelt 1, 2 furnaces of alloy steel with relatively low content of the same element, and clean the furnace lining and ladle used in the previous furnace. Such steel is called brush pot steel. If the brush pot steel is smelted by back blowing method, the content of brushed elements shall be lower than 0.20%~0.50% of the lower limit of the steel specification; if the brush pot steel is smelted by oxidation method, the content of brushed elements shall be lower. In addition, the higher the tapping temperature, the lower the content of brushed elements.

7）刷锅钢种炉料成分的配定原则。在电炉炼钢车间，在冶炼含 Cr、Ni、Mo、W 或 Mn 等高合金钢结束后，接着需冶炼 1 炉或 2 炉含同种元素含量相应较低的合金钢，对上一炉使用的炉衬和钢包进行清洗，这样的钢种被称为刷锅钢种。刷锅钢种如采用返吹法冶炼，被刷元素的含量（质量分数）应低于该钢种规格下限的 0.20%~0.50%；如用氧化法冶炼，被刷元素的含量还要低一些。另外，出钢温度越高的钢种，被刷元素的含量应配得越低。

8) Determination of phosphorus and sulfur. Except for phosphorus and sulfur steel, the lower the content of phosphorus and sulfur in general steel, the better. However, considering the actual situation of steel material, in the batching process, the content of phosphorus and sulfur is less than the value allowed by the process or regulations.

8) 磷、硫的配定。除磷、硫钢外，一般钢中的磷、硫含量均是配得越低越好，但顾及钢铁料的实际情况，在配料过程中，磷、硫含量的配定小于工艺或规程要求所允许的值即可。

9) Matching of aluminum and titanium. In the smelting of electric furnace steel, except for nickel base alloy, the burning loss of aluminum and titanium elements is relatively large, so no matter what method is adopted for smelting, there is generally no artificial blending.

9) 铝、钛的配定。在电炉钢冶炼中，除镍基合金外，铝、钛元素的烧损均较大，因此无论采用何种方法冶炼，一般都不人为配入。

10) Copper distribution. In the smelting process of steel, copper can not be removed, and there is selective oxidation of copper in steel when heated in oxidation atmosphere, which affects the hot working quality of steel. Therefore, the lower the copper content in general steel, the better. Copper in copper steel is mostly added with use.

10) 铜的配定。在钢的冶炼过程中，铜无法去除，且钢中的铜在氧化气氛中加热时存在着选择性的氧化，影响钢的热加工质量。因此，一般钢中的铜含量应配得越低越好，而铜钢中的铜多随用随加。

(2) Proportioning calculation formula:

$$\text{tapping volume} = \text{output} + \text{sprue volume} + \text{steel volume of middle injection pipe} + \text{injection margin}$$

$$\text{output} = \text{standard ingot (Billet) single weight} \times \text{number of pieces} \times \text{relative density coefficient}$$

$$\text{decoction quantity} = \text{single weight of standard decoction} \times \text{number of pieces} \times \text{relative density coefficient}$$

$$\text{steel quantity of middle injection pipe} = \text{single weight of standard middle injection pipe} \times \text{number of pipes} \times \text{relative density coefficient}$$

The injection allowance is the remaining steel water volume after filling the pouring cap mouth, which is generally $0.5\% \sim 1.5\%$ of the steel output. For small capacity, large number of casting disks and small ingot production, the upper limit value is taken; otherwise, the lower limit value is taken.

In the process of batching, the relative density coefficient of steel should not be ignored.

$$\text{Loading capacity} = \frac{\text{Tapping quantity}}{\text{Comprehensive yield of charge}}$$

（2）配料计算公式：

$$出钢量 = 产量 + 汤道量 + 中注管钢量 + 注余量$$
$$产量 = 标准钢锭(钢坯)单重 \times 支数 \times 相对密度系数$$
$$汤道量 = 标准汤道单重 \times 根数 \times 相对密度系数$$
$$中注管钢量 = 标准中注管单重 \times 根数 \times 相对密度系数$$

注余量是浇注帽口充填后的剩余钢水量，一般为出钢量的 0.5%~1.5%。对于容量小、浇注盘数多、生产小锭时，取上限值；反之取下限值。

配料过程中，不可不考虑钢的相对密度系数。

$$装入量 = \frac{出钢量}{炉料综合收得率}$$

The comprehensive recovery rate of charge is determined according to the total amount of impurities and elements in the charge. The larger the burn loss is, the higher the ratio is, the lower the comprehensive recovery rate is.

comprehensive recovery rate of charge = $\sum$ proportioning ratio of various iron and steel materials × recovery rate of various iron and steel materials + $\sum$ addition ratio of various ferroalloys × recovery rate of various ferroalloys

The yield of iron and steel is generally divided into three levels.

The recovery rate of primary steel material is considered as 98%, mainly including returned scrap, mild steel, flat steel, furnace washing steel, forging head, pig iron and intermediate compound surplus material. The surface of this grade steel material is free of rust or less rust.

The recovery rate of secondary steel is 94%, mainly including low-quality steel, railway construction waste equipment, spring steel, wheels, etc.

The yield of the third grade iron and steel material fluctuates greatly, generally considered as 85%~90%, mainly including light and thin miscellaneous iron, chain plate, slag steel, etc. The surface of this grade iron and steel material is seriously rusted, with more dust and impurities.

For the new lining (the first furnace), due to the strong ability of magnesium refractories to absorb iron, the yield of iron and steel materials is lower. Generally, it needs to be equipped with more than 1% of the input.

$$proportioning\ amount = loading\ amount - total\ supplement\ amount\ of$$
$$ferroalloy - iron\ input\ amount\ of\ ore$$

炉料综合收得率是根据炉料中杂质和元素烧损的总量而确定的，烧损越大，配比越高，综合收得率越低。

$$炉料综合收得率 = \sum 各种钢铁料配料比 \times 各种钢铁料收得率 +$$
$$\sum 各种铁合金加入比例 \times 各种铁合金收得率$$

钢铁料的收得率一般分为三级。

一级钢铁料的收得率按98%考虑，主要包括返回废钢、软钢、平钢、洗炉钢、锻头、生铁以及中间合余料等，这级钢铁料表面无锈或少锈。

二级钢铁料的收得率按94%考虑，主要包括低质钢、铁路建筑废器材、弹簧钢、车轮等。

三级钢铁料的收得率波动较大，一般按85%～90%考虑，主要包括轻薄杂铁、链板、渣钢铁等，这级钢铁料表面锈蚀严重，灰尘杂质较多。

对于新炉衬（第一炉），因镁质耐火材料吸附铁的能力较强，钢铁料的收得率更低，一般还需多配装入量的1%左右。

$$配料量 = 装入量 - 铁合金总补加量 - 矿石进铁量$$

iron content of ore = ore content × iron content × iron yield

Generally, the amount of ore added is calculated as 4% of the tapping amount. If the total amount of ferroalloy added is large, the total amount of ferroalloy added shall be deducted from the tapping amount, and then the amount of ore input shall be calculated. The iron content in the ore is about 50%～60%, and the iron recovery rate is considered as 80%. This item is not available for non oxidation smelting because no ore is used.

proportioning amount of various materials = proportioning amount × proportioning ratio of various materials

$$矿石进铁量 = 矿石加入量 × 矿石含铁量 × 铁的收得率$$

矿石的加入量一般按出钢量的4%算，如果铁合金的总补加量较大，需在出钢量中扣除铁合金的总补加量，然后再计算矿石进铁量。矿石中的铁含量为50%～60%，铁的收得率按80%考虑，非氧化法冶炼因不用矿石，故无此项。

$$各种材料配料量 = 配料量 × 各种材料配料比$$

3.1.3 Loading
3.1.3 装料

Charging operation is an important part in the smelting process of electric furnace. It has a great influence on the melting of furnace charge, the burning loss of alloy elements and the service life of furnace lining.

装料操作是电炉冶炼过程中重要的一环，它对炉料的熔化、合金元素的烧损以及炉衬的使用寿命等都有很大的影响。

3.1.3.1 Charging Method
3.1.3.1 装料方法

The most common way of EAF steelmaking is cold charging, which can be divided into manual charging and mechanical charging according to the different ways of steel charging;

mechanical charging can be divided into material trough, hopper and basket charging due to different equipment. At present, it is the top loading of the material basket that is widely used. The charging process is that the furnace charge is installed in the material basket at the bottom of which the chain is used to pin according to certain requirements. When charging, first lift the furnace cover and rotate it to the back of the furnace or pull out the furnace body; then use the crown block to lift the material basket from the furnace top into the furnace, and then pull out the pin to discharge the material into the furnace.

Manual charging is mostly used for electric furnaces with nominal capacity less than 3t. The disadvantages are long charging time, low productivity, large heat loss, high power consumption, high labor intensity, and the briquette and single weight of the furnace charge are limited by the size of the furnace door and the physical strength of people.

Although the charging of material trough or hopper can reduce the labor intensity and make up for some shortcomings of manual charging, the charging time is still long and it is easy to cut and touch the furnace door.

Charging at the top of the charging basket is the most ideal charging method at present. The charging speed of the furnace is fast, which can be completed in 3~5min. The heat loss is small, the electric energy is saved, the service life of the furnace lining can be improved, and the space of the smelting chamber can be fully utilized. In addition, the materials in the basket can be loaded in advance on the material bay or storage yard, with sufficient time and reasonable distribution. The charging position in the basket can still be maintained by the charging in the furnace. If the charging quality is good, the charging can be completed at one time.

电炉炼钢最常见的是冷装料，而冷装按钢铁料的入炉方式不同可分为人工装料和机械装料；机械装料因采用设备不同又分为料槽、料斗、料筐装料等多种。目前，广泛采用的还是料筐顶装料。其装料过程是：将炉料按一定要求装在用铁链销住底部的料筐中。装料时，先抬起炉盖，并将其旋转到炉子的后侧或将炉体开出；然后再用天车将料筐从炉顶吊入炉内，而后拉开销子卸料入炉。

人工装料多用于公称容量小于3t的电炉，缺点是装料时间长，生产率低，热量损失大，电能消耗高，劳动强度大且炉料的块度和单重受炉门尺寸及人的体力限制。

料槽或料斗装料虽能减轻劳动强度，能弥补人工装料的一些缺点，但装料时间仍较长，且易剔碰炉门。

料筐顶装料是目前最理想的装料方法，炉料入炉速度快，只需3~5min就可完成，热量损失小，节约电能，能提高炉衬的使用寿命，还能充分利用熔炼室的空间。另外，料筐中的料可在原料跨间或贮料场上提前装好，时间充裕、布料合理，装入炉内的炉料仍能保持它在料筐中的布料位置，如炉料质量好，一次即可完成装料。

3.1.3.2 Requirements for Loading
3.1.3.2 对装料的要求

In order to shorten the time, ensure the recovery rate of alloy elements, reduce the power consumption and improve the service life of furnace lining, it is required to be accurate, fast in furnace, dense in loading and reasonable in distribution. Pay attention to the following points during operation:

为了缩短时间,保证合金元素的收得率,降低电耗和提高炉衬的使用寿命,装料时要求做到:准确无误,快速入炉,装得致密,布料合理。操作时应注意以下几点:

(1) Prevent wrong installation. First of all, when loading the raw material section, it is required to load the materials strictly according to the batching list, and it is strictly prohibited to load the wrong furnace materials; second, when hanging the basket in front of the furnace, it is necessary to carefully check whether the furnace number, smelting steel type and smelting method of the basket list are consistent with the production plan list in front of the furnace, so as to prevent hoisting the wrong basket.

(1) 防止错装。首先是原料工段装料时,要严格按配料单进行装料,严禁装错炉料;其次是炉前吊筐时,要认真检查随料筐单的炉号、冶炼钢种及冶炼方法等与炉前的生产计划单是否相符,防止吊错料筐。

(2) Fast loading. When the steel is just out of the furnace, the temperature of the furnace is over 1500℃, but the heat dissipation is very fast at this time, and it can be reduced to below 800℃ in a few minutes. Therefore, the preparation before charging should be done well in advance, and the charge should be quickly loaded into the furnace after necessary furnace replenishment, so as to make full use of the residual heat in the furnace, which is of great significance for accelerating charge melting and reducing power consumption.

(2) 快速装料。刚出完钢时,炉膛温度高达1500℃以上,但此时散热很快,几分钟内便可降到800℃以下。因此,应预先做好装料前的准备工作,进行必要的补炉之后快速将炉料装入炉内,以便充分利用炉内的余热,这对于加速炉料熔化、降低电耗等有重要意义。

(3) Reasonable cloth. Reasonable cloth includes the following two meanings:

1) The collocation of various charges should be reasonable. The charge in the furnace shall be dense enough to ensure one-time charging; meanwhile, the conductivity of the charge shall be increased to accelerate melting. Therefore, it is necessary to match large, medium and small materials reasonably. Generally, the material with the weight less than 10kg is small material, the material with the weight of 10~25kg is medium material, and the material with the weight more than 50kg but less than one fiftieth of the total weight of furnace material is large material. According to the production experience, the reasonable proportion is that small material accounts for 15%~20%, medium material accounts for 40%~50% and large material accounts for 40%.

2) The distribution of various charges should be reasonable. According to the characteristics of the temperature distribution in the electric furnace, the reasonable position of all kinds of charge in the furnace, i.e. the basket, is as follows: some small charge is loaded at the bottom,

the amount of which is half of the total amount of small charge, so as to buffer the impact on the furnace bottom during charging, and at the same time, it is conducive to the formation of a molten pool at the bottom of the furnace as soon as possible; then, all the large charge is loaded at the bottom center of the basket, where the temperature is high, which is conducive to the melting of the large charge, and at the same time, it can prevent electricity before the enough liquid steel is deposited at the bottom of the furnace, it will fall to the bottom of the furnace and burn the furnace lining; small materials will be filled between the large materials to ensure the compactness of the furnace materials; medium-sized furnace materials will be installed on and around the large materials; the remaining small materials will be placed on the top so that the electrode can quickly 'penetrate the well' after power transmission, and the arc will be buried in the furnace materials to reduce the thermal radiation of the arc on the furnace cover. If the charge is equipped with pig iron, it shall be installed on the top of the charge or under the electrode to facilitate the use of its carburizing effect to reduce the melting point of the charge and accelerate its melting. If the furnace charge is equipped with alloy, the ferrotungsten and ferromolybdenum with high melting point shall be installed in the high-temperature area around the arc, but not directly below the arc; the volatile ferroalloy such as ferromanganese and nickel plate shall be installed outside the high-temperature area, that is, near the furnace slope, to reduce its volatilization loss; the ferrochromium alloy with high carbon content shall not be directly placed under the electrode.

（3）合理布料。合理布料包括以下两方面的含义：

1）各种炉料的搭配要合理。装入炉内的炉料要足够密实，以保证一次装完；同时，增加炉料的导电性，以加速熔化。为此，必须大、中、小料合理搭配。一般料块质量小于10kg 的为小料，10~25kg 的为中料，大于50kg 而小于炉料总重1/50 的为大料。根据生产经验，合理的配比是小料占15%~20%，中料占40%~50%，大料占40%。

2）各种炉料的分布要合理。根据电炉内温度分布的特点，各种炉料在炉内亦即筐内的合理位置是：底部装一些小料，用量为小料总量的一半，以缓冲装料时对炉底的冲击，同时有利于尽早在炉底形成熔池；然后在料筐的下部中心装全部大料，此处温度高，有利于大料的熔化，同时还可防止电极在炉底尚未积存足够深的钢液前降至炉底而烧坏炉衬；在大料之间填充小料，以保证炉料密实；中型炉料装在大料的上面及四周；最上面放上剩余的小料，以便送电后电极能很快"穿井"，埋弧于炉料之中，减轻电弧对炉盖的热辐射。如果炉料中配有生铁，应装在大料的上面或电极下面，以便利用它的渗碳作用降低大料的熔点，加速其熔化。若炉料中配有合金，熔点高的钨铁、钼铁等应装在电弧周围的高温区，但不能在电弧的正下方；高温下易挥发的铁合金（如锰铁、镍板等）应装在高温区以外，即靠近炉坡处，以减少其挥发损失；容易增碳的铬铁合金也不要直接放在电极下面。

（4）Protect the furnace lining. During charging, the damage of the charge to the furnace lining should also be minimized. Therefore, before charging, a layer of lime with the charge weight of 1.5%~2.0% shall be paved on the bottom of the furnace to relieve the impact of the charge; meanwhile, the lime paved on the bottom of the furnace can also make slag in advance, which is conducive to early phosphorus removal, accelerated temperature rise and gas absorption of the molten steel, etc. When discharging, the distance between the bottom of the basket and the bottom of the furnace shall be as small as possible under the conditions of satisfying the operation, generally about 200~300mm.

（4）保护炉衬。装料时，还应尽量减轻炉料对炉衬的损害。为此，装料前，在炉底上先铺一层为炉料重量1.5%~2.0%的石灰，以缓解炉料的冲击；同时，炉底铺石灰还可以提前造渣，有利于早期去磷、加速升温和钢液的吸气等。卸料时，料筐的底部与炉底的距离在满足操作的条件下尽量小些，一般为200~300mm。

3.1.3.3　Charging and Power Transmission
3.1.3.3　炉料入炉与送电

（1）Charging. The charging at the top of the material basket shall be directed by a specially assigned person. When the furnace cover is pulled out or rotated, the furnace cover shall be fully raised, the electrode shall be raised to the top and the lower end shall be separated from the furnace, so as to prevent the furnace cover or electrode from being damaged. At the same time, the lower end of the electrode shall not exceed the water-cooling ring or insulation ring of the furnace cover, so as to avoid breaking the electrode when shaking and swinging and rolling it to other places, damaging the equipment or injuring people. After the furnace is exposed, the material basket shall be quickly hoisted into the center of the furnace, and shall not be too high, over biased or too low. If it is too high, it is easy to damage the bottom of the furnace, and the crane vibrates greatly; over bias will make the charge layout in the furnace biased, and it is easy to scratch the furnace wall when carrying the basket; too low, it is easy to stick the chain plate of the basket.

（1）炉料入炉。料筐顶装料要有专人指挥，抽炉或旋转炉盖时，炉盖要完全抬起，电极要升到顶点且下端脱离炉膛、以防剐坏炉盖或电极，同时又要求电极下端不许超出炉盖的水冷圈或绝缘圈，避免摇晃摆动时将电极折断而滚落到它处砸坏设备或砸伤人。炉膛裸露后，应迅速将料筐吊入炉内的中心位置，不得过高、过偏与过低。过高容易砸坏炉底，且吊车震动大；过偏将使炉料在炉中布局偏倚，抬料筐时也容易刮擦炉壁；过低易粘坏料筐的链板。

When the operation of steel and slag retention is adopted, more miscellaneous iron shall be padded during loading, and the basket shall be allowed to be raised slightly higher. For multiple charging, the power shall be cut every time. Because there is a large amount of molten steel in the furnace, the charging basket shall be raised higher, which can not only avoid sticking the charging basket, but also reduce any spraying and splashing of flame and molten steel, and also prevent explosion. It is strictly prohibited to put the moist charging into the charging basket for multiple charging.

After the charge is put into the furnace, the over high charge shall be flattened or lifted out to avoid affecting the rotation and closure of the extraction furnace or furnace cover.

采用留钢留渣操作时,装料时应多垫些杂铁,并允许料筐抬得略高些。对于多次装料,每次均要切电,因炉内存有大量的钢液,料筐应抬得再高些,这样既可避免粘坏料筐,又可减少火焰与钢液的任意喷射与飞溅,同时还要防止爆炸,潮湿的炉料严禁装入多次装料的料筐中。

炉料入炉后,对于过高的炉料应压平或吊出,以免影响抽炉或炉盖的旋转与扣合。

(2) Power transmission. After charging into the furnace and before power transmission, the steelmaker and equipment maintenance personnel shall check the furnace cover, electrode, water cooling system, mechanical transmission system, electrical equipment, etc. If any fault is found, it shall be handled in time to avoid shutdown in the smelting process; also check whether the charging is in contact with the furnace door or water cooling system, if any, it shall be removed immediately to avoid breakdown after power transmission.

If the electrode is not long enough, it is better to replace it before power transmission, so as to facilitate the success of one-time drilling. When smelting low carbon high alloy steel, attention shall be paid to the tail or joint of the electrode. If it is found that it is not firm or there is burr, it shall be knocked out to avoid carbon increase in the smelting process. The lower end of the new electrode shall be free of mud or other insulating materials, so as not to affect the arc starting. After the above work is completed and confirmed to be correct, the normal power transmission can be transferred to the melting period.

(2) 送电。炉料入炉后并在送电前,电炉炼钢工和设备维护人员应对炉盖、电极、水冷系统、机械传动系统、电气设备等进行检查,如发现故障要及时处理,以免在冶炼过程中造成停工;还应检查炉料与炉门或水冷系统是否接触,如有接触要立即排除,以免送电后被击穿。

如电极不够长,最好在送电前更换,以利于一次穿井成功。在冶炼低碳高合金钢时应注意电极的接尾或接头,如发现不牢固或有毛刺要打掉,避免冶炼过程中增碳。新换电极下端应无泥土或其他绝缘物质,以免影响起弧。当完成上述工作并确认无误后,方可正常送电转入熔化期。

Exercises

(1) What are the three factors affecting the lining life?

(2) What is the principle of furnace repair?

(3) What are the basic requirements for ingredients?

(4) What are the requirements for loading?

思考题

(1) 影响炉衬寿命的三要素是什么?

(2) 电炉补炉的原则是什么?

(3) 对配料的基本要求是什么?

(4) 装料时对装料的要求是什么?

Task 3.2　Melting Period Operation
任务 3.2　熔化期操作

Mission objectives
任务目标

(1) Understand the physicochemical reaction in melting period.
(1) 了解熔化期的物化反应。
(2) Master the melting period operation.
(2) 掌握熔化期操作。
(3) Master the dephosphorization operation in melting period.
(3) 掌握熔化期脱磷操作。

The main task of the melting period is to melt the solid charge into a uniform liquid with the minimum power consumption on the premise of ensuring the furnace life. At the same time, some physicochemical reactions take place in the furnace, such as removing most of the phosphorus and other impurities in the molten steel, reducing or limiting the gas absorption and element volatilization of the molten steel, etc. In addition, it is another important task of melting period to raise the temperature of melting pool purposefully to create conditions for the next stage of smelting.

熔化期的主要任务是在保证炉体寿命的前提下，以最少的电耗将固体炉料迅速熔化为均匀的液体。在这同时，炉中还伴随着发生一些物化反应，如去除钢液中的大部分磷和其他杂质以及减少或限制钢液的吸气与元素的挥发等。此外，有目的地升高熔池温度，为下一阶段冶炼的顺利进行创造条件，也是熔化期的另一重要任务。

The melting period of traditional EAF is about half of the smelting time of the whole furnace, and the electric energy consumption accounts for 50% ~ 60% of the total electric energy consumption. Therefore, it is of great practical significance to improve the technical and economic indicators of EAF steelmaking by rapidly converting materials and shortening smelting time, which is also another subject that EAF steelmakers have been studying for a long time.

传统的电炉炼钢熔化期约占全炉冶炼时间的一半，电能消耗占总电耗的 50% ~ 60%。因此，快速化料，缩短冶炼时间，对改善电炉炼钢的技术经济指标具有重要的实际意义，这是电炉炼钢工作者长期研究的又一课题。

3.2.1　Melting Process of Charge
3.2.1　炉料的熔化过程

After the start of power transmission, it is the beginning of melting period. The melting process of furnace charge is generally divided into four stages, as shown in Figure 3-1.

送电开始后，就是熔化期的开始，炉料的熔化过程大体上分为如图 3-1 所示的 4 个阶段。

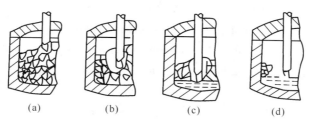

Figure 3-1　Schematic diagram of charge melting process
(a) arc starting stage; (b) well penetrating stage; (c) electrode rising stage;
(d) charge melting stage in low temperature zone

图 3-1　炉料熔化过程示意图
(a) 起弧阶段；(b) 穿井阶段；(c) 电极回升阶段；(d) 低温区炉料熔化阶段

The first stage: arcing stage. After power transmission, the electrode drops. When the electrode end is a certain distance from the furnace charge, due to the effect of strong current, the air in the middle is ionized into ions, and a large number of electrons are released to form a conductive arc, resulting in a large number of light and heat. In the arc starting stage, the time is short, about 3~5min, but the instantaneous short-circuit current often occurs, so the current is generally unstable and causes the impact on the power grid, resulting in the phenomenon of light flashing or television image interference.

第一阶段：起弧阶段。送电后，电极下降，当电极端部距炉料有一定的距离时，由于强大电流的作用，中间的空气被电离成离子，并放出大量的电子而形成导电的电弧，随之产生大量的光和热。起弧阶段的时间较短，为 3~5min，但常出现瞬时短路电流，所以电流一般不稳定并造成了对电网的冲击，从而产生了灯光闪烁或电视图像干扰等现象。

The second stage: through well stage. After arcing, under the action of arc, the charge under the electrode melts first. As the charge melts, the electrode gradually drops and reaches its lowest position, which is the stage of well penetration. Generally speaking, the electric furnace with a large polar circle is often in the center of the charge. The charge is penetrated into three small wells with a diameter of 30%~40% larger than that of the electrode. The charge between the three-phase electrodes of the electric furnace with a small polar circle melts almost at the same time, and it is easy to form a large well at the beginning. In the well penetrating stage, the molten metal droplet under the electrode flows down the gap between the blocks. At the beginning, the furnace temperature is low, and the droplet flows and condenses on the cold material. When the furnace temperature rises, the molten droplet will fall on the furnace bottom and accumulate to form a melting pool and gradually expand.

第二阶段：穿井阶段。起弧后，在电弧的作用下，电极下的炉料首先熔化，随着炉料的熔化，电极逐渐下降并到达它的最低位置，这就是穿井阶段。一般说来，极心圆较大的电炉往往在炉料中央部位，电极把炉料穿成比电极直径大30%~40%的三口小井，而极心圆较小的电炉三相电极间的炉料几乎同时熔化,一开始便容易形成一口大井。在穿井阶段,电极下熔化的金属液滴顺着料块间隙向下流动,开始时炉温较低,液滴边流动边凝结在冷料上,当炉温升高后,熔化的液滴便落在炉底上积存下来形成熔池并逐渐扩大。

The third stage: electrode rising stage. This stage is mainly to melt the charge around the electrode and gradually expand outwards. As the melting continues, the charge in the central part melts, and three small wells form a large well, the pool surface expands and rises, and the electrode also rises correspondingly, which is the stage of electrode recovery. In the process of electrode recovery, the surrounding charge is melted. When only the furnace slope, slag line and other materials near the low temperature zone are left in the furnace, this stage will end.

第三阶段:电极回升阶段。这个阶段主要是熔化电极周围的炉料,并逐渐向外扩大。随着熔化继续进行,中央部分的炉料跟着熔化,三口小井汇合成一口大井,熔池面不断扩大上升,电极也相应向上抬起,这就是电极回升阶段。在电极回升过程中,周围炉料被熔化。当炉内只剩下炉坡、渣线和其他低温区附近的炉料时,该阶段即告结束。

The fourth stage: melting the charge in the low temperature zone. The three-phase arc is similar to the point heat source, the heat radiation of each phase is not uniform, so the temperature distribution in the furnace is not uniform. Generally, the melting rate of the charge is faster at the bottom of the electrode and the hot spot area near the No. 2 electrode, but slower at the low temperature area of the furnace door, both sides of the tapping port and the wall near the No. 1 electrode. The fourth stage is mainly to melt the charge at these parts. In this process, although the electrode continues to pick up slightly, it is not obvious.

第四阶段:熔化低温区炉料阶段。三相电弧近似于点热源,各相的热辐射不均匀,所以炉内的温度分布也不均匀。一般情况下,在电极下边和靠2号电极热点区的炉料熔化较快,而炉门、出钢口两侧及靠1号炉壁处低温区的炉料熔化较慢。第四阶段主要是熔化这些部位的炉料。在此过程中,电极虽然也继续稍有回升,但不明显。

3.2.2 The Main Factors Affecting the Melting of Charge
3.2.2 影响炉料熔化的主要因素

The main energy source of EAF steelmaking is to convert electric energy into heat energy. At present, one of the development trends is to increase the input power of EAF, which is conducive to the melting of furnace burden. Therefore, some high-power and ultra-high-power EAFs have been put into production successively. The second is to use external auxiliary heat sources, such as preheating of furnace burden, oxygen and oxygen combustion nozzle. According to the data, when the preheating temperature of scrap iron before entering the furnace is 500℃, the electric energy can be saved by 1/4, while when the temperature is 600~700℃, the electric energy can be saved by 1/3. If the temperature reaches 900℃, only about 1/2 of the electric energy during cold charging is needed. This means that the melting period will be shortened in proportion to the input power of the transformer. In addition, the hot charge into the furnace can increase the stability of the electric arc, advance the oxygen blowing to aid melting, and shorten the melting period. In order to reduce the consumption of electric energy and accelerate the melting of scrap iron, in addition to timely and reasonable blowing of oxygen to assist melting, oxygen combustion nozzle is used for heating. Oxygen burners use oxygen to support combustion, burning natural gas or light oil, and some use pulverized coal. They are usually used in the dead corner of the low temperature zone in the furnace or at the beginning of melting when the furnace temperature is not high.

电炉炼钢的能源主要是把电能转换成热能,目前发展趋势之一是加大电炉输入功率,从而有利于炉料的熔化,因此一些高功率、超高功率电炉相继投入生产;其次是利用外界辅助热源,如炉料预热、氧气及氧-燃烧嘴等助熔。据资料介绍,废钢铁料入炉前的预热温度为500℃时,可节省电能1/4,而温度为600~700℃时,可节省电能1/3,如果温度达到900℃,只需冷装料时的1/2左右的电能。这就意味着,变压器输入功率不变,熔化期将按相应的比例缩短。此外,热的炉料入炉还可增加电弧的稳定性和提前吹氧助熔,也促使熔化期的缩短。为了减少电能消耗和加速废钢铁料的熔化,在炉内除及时合理地吹氧助熔外,就是利用氧-燃烧嘴加热。氧-燃烧嘴是用氧气助燃,燃烧天然气或轻油,也有的使用煤粉,一般多用于炉内低温区的死角部位或炉温不高的熔化开始阶段。

The melting period is not only directly related to the heat source, but also depends on the use system of electric power, charging method, distribution, chemical composition of furnace charge, smelting process, slag making system and design parameters of furnace body. The higher the useful power of the transformer, the smaller the heat loss of the furnace and the shorter the melting period. In the early stage of melting, because the cold material can absorb a lot of heat, it is advantageous to use high current and the highest voltage in the stage of well penetration and electrode recovery. When the iron in the furnace collapses, the arc light can not be surrounded by the furnace charge. At this time, No. 2 voltage should be replaced appropriately (for ordinary power transformers equipped with 4~6 common secondary voltage). Because the arc light of No. 2 voltage is shorter than that of No. 1 voltage, the short arc light is easy to be surrounded by the slag, so as to reduce the heat consumption caused by the heat radiation, at the same time, it is conducive to the heat conduction in the molten pool, thus shortening the melting of the furnace charge time. Rapid charging can reduce the heat loss and make full use of the residual heat in the furnace to heat the furnace charge. If the material cannot be loaded in time due to the reasons of equal material, manual loading or equipment failure, the melting time will be prolonged. Reasonable material is also one of the effective measures to shorten the melting period. For example, if the charge is loose or a large refractory low carbon scrap is loaded in the upper part, the charge bridge is easy to form in the furnace during melting, and the melting time is easy to be prolonged. If the electrode is interrupted due to iron collapse again, the melting time is even longer. The melting rate also depends on the chemical composition of the charge. The resistance of carbon steel with carbon content of 0.20% is 6 times of that of copper, while the resistance of carbon steel with carbon content of 0.90% is 14 times of that of copper; the charge can be regarded as the secondary circuit of current, according to Joule's Law:

$$Q = I^2 Rt$$

熔化期的长短不仅与热源有直接关系，而且还取决于电力使用制度、装料方法、布料情况、炉料的化学成分、冶炼工艺、造渣制度及炉体的设计参数等。变压器的有用功率越大，炉子的热损失越小，熔化期就越短。在熔化初期，由于冷料能够吸收大量的热量，因此在穿井和电极回升阶段，使用大电流和最高级电压是有利的。当炉中塌铁后，弧光不能被炉料包围，这时应更换 2 号电压较为合适（对备有 4~6 个常用的次级电压的普通功率变压器而言），因 2 号电压的弧光较 1 号电压的短，短的弧光容易被熔渣包围，这样就减少了热辐射造成的热耗，同时也有利于熔池内的热传导，从而缩短了炉料的熔化时间。快速装料能减少热损失，充分利用炉中的余热来加热炉料。如果等料、人工装料或因设备损坏等原因不能及时装料，势必延长熔化时间。合理的布料也是缩短熔化期的有效措施之一。如炉料装得疏松或在上部装入大块难熔的低碳废钢等，在熔化时炉内容易形成料桥，极易延长熔化时间，如再出现塌铁而把电极打断，更拖延了熔化时间。熔化速度也取决于炉料的化学成分。碳含量（质量分数）为 0.20% 碳素钢的电阻是铜的电阻的 6 倍，而碳含量（质量分数）为 0.90% 碳素钢的电阻是铜的 14 倍；炉料可看成是电流的二次线路，根据焦耳定律：

$$Q = I^2 Rt$$

The greater the resistance of the charge is, the more heat will be generated when the current passes through. Moreover, the melting point of high carbon steel is lower than that of low carbon steel, so high carbon materials melt faster than low carbon materials. In addition, the easily oxidized elements such as Si, Al, P in the charge can react exothermically with oxygen, and the reaction heat can be regarded as the auxiliary heat source of charge melting. Therefore, the higher the content of easily oxidized elements in the charge, the better the charge melting. The melting time of the charge can also be shortened by the reasonable expansion of the charge. Early slagging helps to melt the charge. Early slagging can not only prevent the intake of molten steel, but also reduce the loss of heat. If the foam slag is submerged arc operation, it can not only reduce the thermal radiation of arc to the furnace wall, but also facilitate the heating and heating of the molten pool. With the operation of steel slag retention, oxygen blowing can be used to aid melting after charging, and the melting period can be shortened obviously. The design parameters of the furnace body also have a direct impact on the melting of the furnace charge. When the diameter of the furnace is larger than the diameter of the polar circle, the charge melts more slowly; for example, when the diameter of the furnace is smaller than the diameter of the polar circle, the charge melts faster. For the furnace body with deep melting pool and small furnace diameter, the melting speed of the charge is also relatively fast because of the small heat dissipation surface, the long travel of the electrode and the fast melting speed of the charge.

炉料的电阻越大，电流通过时所产生的热量越多。更何况高碳钢的熔点比低碳钢的熔点低，所以高碳料比低碳料熔化得快。另外，炉料中含 Si、Al、P 等易氧化元素能与氧发生放热反应，反应热可看成是炉料熔化的辅助热源，因此炉料中易氧化元素的含量越高越有利于炉料的熔化。金属炉料的合理扩装也可缩短炉料熔化时间。早期造渣有助于炉料的熔化。早期造渣不仅可以防止钢液的吸气，同时也能减少热量的散失，如果采用泡沫渣埋弧操作，不仅可以减轻弧光对炉壁的热辐射，而且更有利于熔池的加热与升温。采用留钢留渣操作，在装料后就可吹氧助熔，也可使熔化期明显缩短。炉体的设计参数对炉料的熔化也有直接影响。当炉膛的直径与极心圆直径之比较大时，炉料熔化得较慢；如炉膛的直径与极心圆直径之比较小时，炉料熔化得就快。对于熔池较深、炉膛直径略小的炉体，因散热面小，电极所走的行程长，炉料的熔化速度也是比较快的。

3.2.3 Physicochemical Reaction in Melting Period
3.2.3 熔化期的物化反应

At the same time of charge melting, various physicochemical reactions take place in the molten pool, mainly including element volatilization and oxidation, liquid steel suction, heat transfer and dissipation, and inclusion floatation.

炉料熔化的同时，熔池中也发生各种各样的物化反应，主要有元素的挥发和氧化、钢液的吸气、热量的传递与散失以及夹杂物的上浮等。

3.2.3.1 Volatilization of Elements
3.2.3.1 元素的挥发

The melting of charge is accompanied by the partial volatilization of elements. There are two forms of volatilization: direct volatilization and indirect volatilization. Direct volatilization occurs when the temperature exceeds the boiling point of the element. The temperature of the arc is as high as 4000~6000℃, and the boiling point of the most refractory element W is only 5900℃. Zn and Pb with low boiling point are more likely to volatilize. Indirect volatilization is carried out by the oxides of the elements, that is, the oxides are formed first, and then the oxides are volatilized at high temperature. Generally speaking, the boiling point of most metal oxides is lower than that of the metal. For example, the boiling point of Mo is 4800℃, while MoO_3. As the boiling point of 1100℃, the volatilization of many metal oxides is often prior to the direct volatilization of the element. There are many metal oxides in the smoke and dust escaping from the furnace door or electrode hole during the melting period, the most of which is Fe_2O_3. This is because the proportion of iron in the furnace charge is the largest, and the vapor pressure of liquid iron is also large, so the smoke and dust escaping during the melting period are mostly brownish red.

炉料熔化的同时，伴随着元素的部分挥发。挥发有直接挥发和间接挥发两种形式。直接挥发是因温度超过元素的沸点而产生的。电弧的温度高达 4000~6000℃，而最难熔元素 W 的沸点也仅为 5900℃，至于低沸点的 Zn、Pb 等就更容易挥发了。间接挥发是通过元素的氧化物进行的，即先形成氧化物，然后氧化物在高温下挥发逸出。一般说来，多数金属氧化物的沸点低于该金属的沸点，如 Mo 的沸点为 4800℃，而 MoO_3 的沸点仅为 1100℃，因此许多金属氧化物的挥发往往先于该元素的直接挥发。熔化期从炉门或电极孔逸出的烟尘中含有许多金属氧化物，其中最多的还是 Fe_2O_3，这是因为铁在炉料中占的比例最大，液态铁的蒸气压也较大，所以熔化期逸出的烟尘多为棕红色。

3.2.3.2 Oxidation of Elements
3.2.3.2 元素的氧化

In addition to the volatilization of elements, there is also oxidation of elements during the melting of furnace charge. This is due to the existence of oxygen sources in the furnace: first, rust on the surface of furnace charge; second, furnace gas; third, ore added for dephosphorization or oxygen introduced for melting aid. In the process of charge melting, the amount of element oxidation loss is related to the characteristics, content, smelting method, charge surface quality and oxygen blowing strength (pressure, flow, time) and other factors. The oxidation loss of Fe, C and Mn is similar in the oxidation method and the back blowing method. In general, Al, Ti and Si elements are oxidized almost completely in the oxidation process, while P can only be oxidized mostly. However, in the back blowing process, the oxidation loss of these elements is slightly less due to the lack of ore remelting, and the least in the non oxidation process. When smelting high alloy steel, for example, the amount of Si in charge is more than 1.0%, and the amount of oxidation loss of Si is about 50% ~ 70%. The oxidation loss of iron is usually 2% ~ 6%. The worse the quality of scrap is, the longer the melting time is, the stronger the oxygen blowing strength is, and the greater the oxidation loss of iron is. The oxidation loss of carbon is generally 0.60%, but the loss of carbon is not too large when oxygen is not used. When oxygen is used, the change of carbon is related to carbon content and oxygen blowing strength. When the amount of carbon in the charge is less than 0.30%, the oxidation loss of carbon is small and can be made up by electrode carburization; when the amount of carbon is more than 0.30%, the oxidation loss of carbon is more. The larger the oxygen pressure, the more the flow rate and the longer the oxygen blowing time are, the more the carbon loss will be. The oxidation loss of carbon also decreases with the increase of silicon content in the charge, which is due to the fact that the affinity of silicon with oxygen is greater than that of carbon when the temperature is below 1530℃. In the melting process of furnace charge, sometimes the molten pool boiling is caused by iron collapse, which will also increase the oxidation loss of carbon. This is mainly because the molten metal in the molten pool is not covered by slag, and a large amount of FeO is accumulated on the liquid surface to react with carbon.

炉料熔化时，除产生元素的挥发外，还存在着元素的氧化。这是因炉中存在着氧的来源：一是炉料的表面铁锈；二是炉气；三是为了脱磷而加入的矿石或为了助熔而引入的氧等。在炉料熔化过程中，元素氧化损失量与元素的特性、含量、冶炼方法、炉料表面质量及吹氧强度（压力、流量、时间）等因素有关。Fe、C、Mn的氧化损失量在氧化法和返吹法中基本相似。在一般情况下，Al、Ti、Si元素在氧化法中几乎全部氧化掉，P只能大部分氧化，但这些元素在返吹法中，因不使用矿石助熔，氧化损失略少些，而在不氧化法中为最少。在冶炼高合金钢时，如炉料的配$w(Si)>1.0\%$，Si的氧化损失量为50% ~ 70%。铁的氧化损失通常为2% ~ 6%。废钢质量越差，熔化时间越长，吹氧强度越大，铁的氧化损失也越大。碳的氧化损失量一般为0.60%，但不用氧时碳的损失不太大。而用氧时，碳的变化与钢液中的碳含量、吹氧强度有关。当炉料中的 $w(C)<0.30\%$ 时，C 的氧化损失不大，并可为电极增碳所弥补；$w(C)>0.30\%$ 时，C 的氧化损失要多些。吹氧助熔的氧气压力越大、流量越多、吹氧时间越长，C 的损失也越多。C 的氧化损失还随炉料中 Si 含量的增高而降低，这是因为 Si 同 O 的亲和力在 1530℃ 以下时大于碳的缘故。炉料熔化过程中，有时因塌铁而引起熔池沸腾，也会使 C 的氧化损失增加，这主要是由于熔池中的金属液无熔渣覆盖，液面富集大量的 FeO 与 C 反应的结果。

3.2.3.3 Liquid Steel Suction
3.2.3.3 钢液的吸气

In general, the solubility of gas in molten steel increases with the increase of temperature, and the hydrogen and nitrogen decomposed by high temperature arc will be dissolved in molten steel directly or through slag layer due to the increase of temperature. In the melting period, the liquid steel has a good suction condition, because in addition to the atmosphere, there is a certain amount of moisture in the charge. In addition, the molten steel droplet in the early stage of melting is exposed when it moves downward, while the molten pool in the early stage sometimes has no slag cover, and the droplet directly contacts with the furnace gas. In order to reduce the suction of molten steel, the molten slag should be made as early as possible. The gas content in the steel can also be reduced by oxygen blowing and melting aid with reasonable melting period.

在一般情况下，气体在钢液中的溶解度随温度的升高而增加，被高温电弧分解出的氢和氮会因温度的升高直接或通过渣层溶解于钢液中。在熔化期，钢液具有较好的吸气条件，这是因为除大气外，炉料中还含有一定的水分。而且熔化初期的钢液液滴向下移动时是裸露的，而初期的熔池有时又无熔渣覆盖，液滴直接与炉气接触。为了减少钢液的吸气量，应尽早造好熔化渣。熔化期合理的吹氧助熔也能降低钢中的气体含量。

3.2.3.4 Heat Transfer and Loss
3.2.3.4 热量的传递与散失

Heat transfer and loss belong to physical process. During the melting period, heat conduction is mainly carried out in the molten pool. In addition to absorbing the residual heat of the furnace lining, most of the heat is obtained from the arc, and the heat is transferred from the top to the bottom during the arc starting and well penetrating stages; when the molten pool is covered with slag, the heat is transferred to the molten steel through the slag, and the heat is still transferred from the top to the bottom, generally speaking, the temperature of the slag is higher than that of the molten steel. Of course, there is radiation and reflection of heat in the furnace, but not the main one. As for the auxiliary heat source, the transfer direction is different due to the different ways provided. After the molten pool appears, if there is no slag cover or less slag at the initial stage, the heat loss is serious. In order to reduce heat dissipation, molten slag should be made as early as possible.

热量的传递与散失属于物理过程。熔化期熔池中主要进行着热传导。炉料除了吸收炉衬的余热外，绝大部分热量是从电弧获取的，在起弧和穿井阶段热量由上向下传递；当熔池有熔渣覆盖后，热量通过熔渣传给钢液，这时的热量仍是由上向下传递的，一般来说，熔渣的温度高于钢液的温度。当然，在炉中还有热量的辐射与反射，但不是主要的。关于辅助热源，由于提供的方式不同，传递方向也不同。熔池出现后，初期如无熔渣覆盖或熔渣较少，热量散失严重。为了减少散热，应尽早造好熔化渣。

3.2.3.5 Floatation of Non-metallic Inclusions during Melting
3.2.3.5 熔化期非金属夹杂物的上浮

After the molten pool appears, there are internal and external inclusions in the molten steel. With the expansion of the molten pool, these inclusions will float up in varying degrees, and they are one of the sources of molten slag. It has been proved that reasonable oxygen blowing and slag making can make the inclusions float up. After oxygen blowing, due to the effect of oxygen flow, the molten pool is partially boiling, which is helpful to the collision and floatation of inclusions; the ideal molten slag is not only beneficial to dephosphorization, but also can capture and absorb non-metallic inclusions well.

熔池出现后，钢液中就存在着内在夹杂和外在夹杂，随着熔池的扩大，这些夹杂物也就有不同程度的上浮，它们是熔化渣的来源之一。实践证明，合理的吹氧助熔和尽早造好熔化渣能促使夹杂充分上浮。吹氧后，由于氧气流的作用，造成熔池局部沸腾，进而有助于夹杂物的碰撞和上浮；理想的熔化渣不仅对脱磷有利，而且还能很好地捕捉、吸附非金属夹杂物。

3.2.4 Dephosphorization Operation in Melting Period
3.2.4 熔化期脱磷操作

The correct operation in the melting period can remove 50%~70% of the phosphorus in the steel, and the remaining phosphorus can be removed in the oxidation period by means of the interface reaction between slag and steel, automatic slag flow, new slag making or powder spraying dephosphorization. Therefore, a mature EAF steelmaker should grasp the dephosphorization operation tightly in the melting period.

In the melting period, the molten slag should be prepared in advance, with proper basicity and good fluidity, it can create favorable conditions for early dephosphorization. In addition, when conditions permit, in addition to adding the fusible aid ore, the iron oxide scale or ore powder with a material weight of 1% can be added in batches during the majority of melting, or a small amount of iron ore can be added at the same time of lining the bottom ash of the furnace, so as to improve the oxidation ability of the melting slag; after the majority of the furnace material is melted or fully melted, the partial melting slag can be removed, and for the steel with high phosphorus furnace material or strict phosphorus specification requirements, it is also possible to add the iron oxide scale or ore powder with a material weight of 1% in batches during the majority of melting. It is an effective way to strengthen dephosphorization to remove all the phosphorus and make new slag. At this time, the dephosphorization efficiency can reach 50% ~70%, and the remaining phosphorus in the molten steel will be transferred to the initial stage of oxidation for further treatment.

熔化期的正确操作,可以将钢中的磷去除50%~70%,剩余的残存磷在氧化期借助于渣钢间的界面反应、自动流渣、补造新渣或采用喷粉脱磷等办法继续去除。因此,一个成熟的电炉炼钢工,应在熔化期紧紧地抓住脱磷操作。

熔化期提前造好熔化渣,并使之具有适当的碱度和较好的流动性,能为前期脱磷创造有利的条件。另外,在条件允许的情况下,除加入助熔矿石外,还可在大半熔时分批加入料重1%的氧化铁皮或矿石粉,或在垫炉底灰的同时装入少量的铁矿石等,从而提高熔化渣的氧化能力;在炉料大半熔或全熔后扒除部分熔化渣,对于高磷炉料或磷规格要求较严的钢种,也可全部扒除,然后重造新渣,更是强化脱磷的行之有效的好办法,此时去磷效率可达50%~70%,而钢液中的剩余磷移到氧化初期去继续处理。

3.2.5 Melting Period Operation
3.2.5 熔化期操作

After power transmission, the furnace door shall be closed tightly, the tapping port shall be closed, and the joint between the furnace cover and the furnace wall and the charging hole shall be fastened to prevent cold air from entering the furnace. At the end of the arc starting stage, the length of the discharge electrode should be adjusted to ensure the success of a well penetration and to meet the needs of the whole furnace smelting. The furnace equipped with oxygen burner device shall also be ignited timely to make the furnace charge melt synchronously. When it is necessary to load several times, it shall be loaded when the smelting chamber can accommodate the material in the next basket after each iron collapse. In the melting process of the furnace charge, oxygen blowing, iron pushing or ore adding should be carried out timely, as well as early slag making and dephosphorization. At the end of melting, if it is found that the fully melted carbon can not meet the process requirements, the carburization operation should be carried out first.

送电后应紧闭炉门,堵好出钢口,扣严炉盖与炉壁的接合处及加料孔等,以防冷空气进入炉内。在起弧阶段结束后,还要调放电极长度,使一次穿井成功并能保证全炉冶炼的需要。备有氧-燃烧嘴装置的炉子也应适时点燃,以使炉料能够同步熔化。需多次装料时,在炉料每次塌铁后,熔炼室能容纳下一料筐中的料时再装入。在炉料熔化过程中,还应适时地进行吹氧、推铁或加矿助熔及早期造渣与脱磷等操作。熔化末期如果发现全熔碳不能满足工艺要求,一般应先进行增碳操作。

Generally, the amount of molten slag is 2% ~3% of the material weight, and the higher the furnace power is, the higher the upper limit value is. After the charge is fully melted and stirred, take the full analysis sample, then remove part of the melted slag and make up new slag. If it is difficult to dephosphorize or if a large amount of MgO is found in the slag, it can also be completely slagged and remade. When the bath temperature rises to meet the process requirements, it can be transferred to the next stage of smelting.

熔化渣的渣量一般为料重的2%~3%，电炉功率越高越取上限值。炉料全熔并经搅拌后，取全分析样，然后扒除部分熔化渣，补造新渣。如果认为脱磷困难或发现熔渣中含有大量的 MgO，也可进行全扒渣，重新造渣。当熔池温度升到符合工艺要求时，方可转入下一阶段的冶炼。

Exercises

(1) What are the four stages of charge melting?

(2) What are the main factors affecting the melting of furnace charge?

(3) What are the physicochemical reactions in the melting period?

思考题

(1) 炉料熔化包括哪四个阶段？

(2) 影响炉料熔化的主要因素是什么？

(3) 熔化期的物化反应都有哪些？

Task 3.3　Operation in Oxidation Period
任务3.3　氧化期及操作

Mission objectives

任务目标

(1) Understand oxidation methods.

(1) 了解氧化方法。

(2) Master the operation of oxidation period.

(2) 掌握氧化期操作。

(3) Master the operation of dephosphorization and decarbonization in oxidation stage.

(3) 掌握氧化期脱磷脱碳操作。

At present, the oxidation period is mainly to control smelting temperature, and to supply oxygen and decarbonization as a means to promote the pool boiling, and quickly complete the designated tasks. At the same time, it also creates favorable conditions for reduction refining.

目前，氧化期主要是以控制冶炼温度为主，并以供氧和脱碳为手段，促进熔池激烈沸腾，迅速完成所指定的各项任务。在这同时，也为还原精炼创造有利的条件。

The main tasks of oxidation period of electric furnace without external refining are as follows:

(1) Continue and finally complete the dephosphorization task of the molten steel, so that the phosphorus in the steel can be reduced to the allowable content range specified in the regulations;

(2) Remove gas from liquid steel;

(3) Remove non-metallic inclusions from molten steel;

(4) Heating and homogenizing the temperature of molten steel to meet the process requirements, generally to reach or higher than the tapping temperature, to create conditions for the refining of molten steel.

不配备炉外精炼的电炉氧化期的主要任务如下:
(1) 继续并最终完成钢液的脱磷任务,使钢中磷降到规程规定的允许含量范围内;
(2) 去除钢液中的气体;
(3) 去除钢液中的非金属夹杂物;
(4) 加热并均匀钢液温度,使之满足工艺要求,一般是达到或高于出钢温度,为钢液的精炼创造条件。

At the same time of the above tasks, the C, Si, Mn, Cr and other impurities in the molten steel are oxidized to varying degrees. The electric furnace is only a high-efficiency melting, dephosphorization and heating tool for smelting equipped with refining device outside the furnace. Under this condition, the removal of gas and non-metallic inclusions in the molten steel is carried out outside the furnace, and the task of oxidation period is reduced.

在上述任务完成的同时,钢液中的C、Si、Mn、Cr等元素及其他杂质也发生不同程度的氧化。配备炉外精炼装置的冶炼,电炉只是一个高效率的熔化、脱磷与升温的工具。在这种条件下,钢液中的气体及非金属夹杂物的去除等,均移至炉外进行,而氧化期的任务也就得以减轻。

3.3.1 Oxidation Method
3.3.1 氧化方法

3.3.1.1 Ore Oxidation
3.3.1.1 矿石氧化法

Ore oxidation is an indirect way of oxygen supply. It mainly uses the oxygen in iron ore or other metallized ore to realize the oxidation of C, Si, Mn and other impurities in molten steel through diffusion transfer.

The characteristics of this method are high concentration of FeO in slag and good dephosphorization effect. The reaction of carbon and [FeO] is the main reaction in the decarburization process, but [FeO] must be realized through the diffusion transfer of (FeO), so the decarburization speed is slow and the oxidation time is long. The decomposition of iron ore is endothermic reaction, which will reduce the temperature of molten pool, so the smelting temperature should be high enough before ore is added into the furnace. It is easy to bring other inclusions into the molten steel of ore oxidation process. Because of the high content of FeO in slag, the fluidity of slag is better.

矿石氧化法属于间接方式的供氧,它主要是利用铁矿石或其他金属化矿石中的氧通过扩散转移来实现钢液中的C、Si、Mn等元素及其他杂质的氧化。

该法的特点是渣中(FeO)浓度高,脱磷效果好。碳和[FeO]的反应是脱碳过程的主要反应，但［FeO］必须通过（FeO）的扩散转移来实现，因此脱碳速度慢，氧化时间长。而铁矿石的分解是吸热反应，会降低熔池温度，所以矿石加入前炉中应具有足够高的冶炼温度。矿石氧化法的钢液中容易带进其他夹杂。因渣中（FeO）含量高，所以熔渣的流动性较好。

3.3.1.2 Oxygen Oxidation
3.3.1.2 氧气氧化法

Oxygen oxidation is also called pure oxygen oxidation. It mainly uses the direct action of oxygen, C, Si, Mn and other impurities in steel to complete the oxidation of molten steel. In addition, the following reactions occur in the molten pool after oxygen blowing:

$$O_2 + 2[Fe] = 2[FeO]$$
$$[FeO] = (FeO)$$

氧气氧化法又称纯氧氧化法。它主要是利用氧气和钢中的 C、Si、Mn 等元素及其他杂质的直接作用来完成钢液的氧化。除此之外，吹氧后，熔池中还发生下述反应：

$$O_2 + 2[Fe] = 2[FeO]$$
$$[FeO] = (FeO)$$

There are essential differences between oxygen oxidation and ore oxidation. During oxygen oxidation, due to the direct effect of pure oxygen on the molten steel, the kinetic conditions of each element oxidation are good. In the case of high oxygen supply intensity, it is more conducive to the smelting of low carbon steel or ultra-low carbon steel.

Oxygen oxidation belongs to exothermic reaction, which is beneficial to increase and even bath temperature and reduce electric energy consumption. In addition, after oxygen oxidation, the molten steel is pure and less impurities are brought in, and after oxygen blowing, the oxygen content in the molten steel is also less, so it is conducive to the deoxidization of the later molten steel. However, due to the low content of FeO, the dephosphorization effect is poor and the fluidity of slag is poor.

氧气氧化和矿石氧化存在着本质的不同。氧气氧化时，由于纯氧对钢液的直接作用，各元素氧化的动力学条件好，在供氧强度较高的情况下，更有利于低碳钢或超低碳钢的冶炼。

氧气氧化属于放热反应，进而也有利于提高和均匀熔池温度而减少电能消耗。此外，氧气氧化后，钢液纯洁，带进其他杂质少，且吹氧后，钢液中的氧含量也少，所以又有利于后步钢液的脱氧。但由于（FeO）含量不高，因此脱磷效果差，熔渣的流动性也差。

3.3.1.3 Comprehensive Oxidation of Ore and Oxygen
3.3.1.3 矿、氧综合氧化法

In the process of EAF steel production, ore oxidation and oxygen oxidation are often used alternately or at the same time, which is called comprehensive oxidation of ore and oxygen. It is characterized by decarburization and rapid temperature rise, which not only does not affect the dephosphorization of molten steel, but also significantly shortens the smelting time. However, if the method is not skilled, it is difficult to control the final decarburization accurately.

在电炉钢生产过程中，矿石氧化和氧气氧化经常交替穿插或同时并用，这就是所谓的矿、氧综合氧化。其特点是脱碳、升温速度快，既不影响钢液的脱磷，又能显著缩短冶炼时间。但该法如不能熟练操作，便难以准确地控制终脱碳。

3.3.2 Decarbonization Operation
3.3.2 脱碳操作

3.3.2.1 Ore addition and Decarbonization of Molten Steel
3.3.2.1 钢液的加矿脱碳

The total process of decarburization is endothermic, because the melting and decomposition of ore and the diffusion and transfer of FeO are endothermic. At the beginning of adding ore and decarbonizing, the temperature of the molten steel must be high enough, generally higher than 1530℃. In order to avoid the rapid cooling of the molten pool, the ore should be added in batches, the amount of each batch is about 1.0% ~ 1.5% of the weight of the molten steel, and when the reaction of the previous batch of ore begins to weaken, the next batch of ore should be added, with an interval of 5 ~ 7min. The uniform and intense boiling of the molten pool is mainly controlled by the addition speed of the ore and keeping the appropriate interval time. When the temperature of the molten pool is high, the addition speed of the ore can not be too fast. If there is a strong flame at the furnace door and electrode hole, the ore addition shall be stopped to avoid splashing or steel running accidents.

由于矿石的熔化与分解及 FeO 的扩散转移均吸热，所以脱碳反应的总过程是吸热。钢液的加矿脱碳开始时必须要有足够高的温度，一般应大于1530℃。为了避免熔池急剧降温，矿石应分批加入，每批的加入量为钢液重量的 1.0% ~ 1.5%，而在前一批矿石反应开始减弱时，再加下一批矿石，间隔时间为 5~7min。熔池的均匀激烈沸腾主要通过对矿石的加入速度和保持合适的间隔时间来控制，当熔池温度较高时，矿石的加入速度也不能太快，如在炉门及电极孔冒出猛烈的火焰，则应停止加矿，以避免发生喷溅或跑钢事故。

In principle, the ore addition and decarbonization of molten steel are carried out under high temperature and thin slag. However, in consideration of the continuous dephosphorization and temperature rise of the molten steel, the temperature control is slow first and then fast, the slag volume is large first and then thin, and there should be enough alkalinity and good fluidity. The viscous slag is not only unfavorable to dephosphorization, but also unfavorable to the diffusion of FeO and the removal of CO bubbles. Especially when the temperature of molten steel is not too high, the molten pool is easy to be 'silent'. The molten pool does not boil after ore addition. At this time, the ore addition should be stopped immediately, and fluorite should be used to adjust the fluidity of the slag and raise the temperature.

钢液的加矿脱碳原则上是在高温、薄渣下进行的。但考虑到钢液的继续脱磷与升温，温度控制是先慢后快，渣量是先大后薄，且还要有足够的碱度及良好的流动性。黏稠的熔渣不仅不利于脱磷，也不利于（FeO）的扩散及 CO 气泡的排除，特别是在钢液温度不太高的情况下，熔池容易出现"寂静"的现象，加矿后熔池不沸腾，这时应立即停止加矿，应用萤石调整熔渣的流动性并升温。

In the early stage of decarburization, the slag with good fluidity is foamed under the action of CO bubbles and automatically flows out through the furnace door. If it can not flow out, it should be adjusted, otherwise it will be difficult to remove carbon from high temperature and thin slag. In order to speed up the melting and decomposition of ore, but also to reduce the temperature of molten pool, when conditions permit, ore should be baked at a high temperature of more than 800℃ for 4 hours before use. The amount of ore added is determined by the amount of decarburization. Theoretical calculation and experience show that for every 0.01% of oxidized C, about 1kg ore is used per ton of molten steel. The ideal full melting carbon should meet the requirements of the process, but sometimes the amount of decarburization is too large or insufficient due to the loading delay or improper melting aid. Too much decarburization not only increases the consumption of various raw materials, but also prolongs the smelting time. If the decarburization is insufficient, it needs to increase the carbon. These two situations are not good for operation and should be avoided as far as possible.

脱碳初期，流动性良好的熔渣在 CO 气泡的作用下呈泡沫状，并经炉门能自动流出，如不能流出应进行调整，否则以后也难以做到高温、薄渣脱碳。为了加速矿石的熔化与分解，且又不过多地降低熔池温度，当条件允许时，矿石应预先在大于 800℃ 的高温下烘烤 4h 后使用。矿石的加入量由脱碳量决定，理论计算及经验表明，每氧化碳 0.01%，每吨钢液约用矿石 1kg。理想的全熔碳应满足工艺的要求，但因装料贻误或助熔不当，有时出现脱碳量过大或不足。脱碳量过大不仅增加了各种原材料的消耗，而且也延长冶炼时间，脱碳量不足需进行增碳，这两种情况对操作不利，应尽量避免。

3.3.2.2 Oxygen Blowing Decarburization of Molten Steel
3.3.2.2 钢液的吹氧脱碳

There are two kinds of decarburization of molten steel: direct oxidation and indirect oxidation. The direct reaction between the oxygen blown into the molten steel and the carbon in the molten steel belongs to the direct oxidation of carbon, while the oxygen blown into the molten steel first reacts with the iron in the molten steel, and then the generated [FeO] reacts with the carbon in the molten steel, which belongs to the indirect oxidation of carbon.

钢液的吹氧脱碳有碳的直接氧化和碳的间接氧化两种情况。吹入钢液中的氧直接与钢液中的碳发生反应属于碳的直接氧化，而吹入钢液中的氧先与钢液中的铁反应，然后生成

的 [FeO] 再与钢液中的碳进行反应，属于碳的间接氧化。

The high pressure oxygen flow blown into the molten steel catches the carbon around the bubble in the form of a large number of dispersed bubbles, and reacts on the bubble surface. At the same time, [FeO] formed around the oxygen bubble interacts with carbon in the molten steel, and the reaction products also enter the bubble. The appearance and diffusion of [FeO] increase the oxygen content in the molten steel, so the oxidation of carbon can be carried out not only in the place where oxygen is blown directly, but also in other parts of the molten pool. The biggest characteristic of oxygen blowing decarburization is that the decarburization speed is fast, generally about (0.03~0.05)%/min, and the higher the liquid steel temperature, the greater the oxygen supply, the higher the carbon content in the steel, the faster the decarburization speed.

吹入钢液中的高压氧气流以大量弥散的气泡形式在钢液中捕捉气泡周围的碳，并在气泡表面进行反应。与此同时，氧气泡周围形成的 [FeO] 与钢液中的碳作用，反应产物也进入气泡中。而 [FeO] 的出现与扩散，又提高了钢液中的氧含量，因此碳的氧化不仅可以在直接吹氧的地方进行，而且也能在熔池中的其他部位进行。吹氧脱碳最大的特点是脱碳速度快，一般为 (0.03~0.05)%/min，而且钢液温度越高、供氧量越大、钢中的碳含量越高，脱碳速度越快。

When the carbon content in the molten steel is reduced to less than 0.10%, the oxygen required for carbon balance in the molten steel will rise sharply, and the oxygen required for carbon balance in the molten steel will also rise. In order to maintain the decarburization speed, it is necessary to increase the oxygen supply. The decarburization of ore adding is affected by the decrease of furnace temperature, which can not be added too much at a time, and the diffusion transfer of FeO into the steel is the limiting ring. Therefore, when the carbon content in the molten steel drops below 0.10%, the decarburization by blowing is better than that by adding ore, and the speed of the two is also significantly different. The production practice also proves that when smelting low carbon or ultra-low carbon steel, oxygen blowing is easy to reduce the carbon to a very low level, and the oxidation loss of alloy elements is less than that of ore, which makes it possible to smelting high alloy steel by back blowing method and recover the precious alloy elements in furnace burden. Under the same conditions, the content of FeO in slag and the final content of FeO in molten steel are less, which can reduce the deoxidization burden of molten steel refining. However, dephosphorization conditions have deteriorated, so dephosphorization task must be completed at the end of melting or the beginning of oxidation and when the temperature of molten steel is not too high. The smelting time of oxygen blowing decarburization is short, which can increase the output by more than 20%, reduce the power consumption by 15%~30%, the electrode consumption by 15%~30%, the total cost by about 6%~8%, and improve the quality of steel. When blowing oxygen and reducing carbon, it is better to choose a higher oxygen pressure. Because of the high oxygen pressure, the oxygen flow can be blown into deeper parts in the molten steel, and can split into more small bubbles, thus improving the utilization rate of oxygen. In addition, high oxygen pressure can also reduce the consumption of oxygen pipe, which is due to the increase of decarburization speed, the reduction of oxygen blowing time, the increase of oxygen flow rate, and the enhancement of cooling effect on the pipe wall end.

当钢液中的碳含量（质量分数）降低到 0.10% 以下时，钢液中所需与碳平衡的氧量将急剧上升，而与钢液中碳平衡所需渣中的氧量也是上升的，这时要保持脱碳速度，就必须增加供氧量，加矿脱碳受到炉温下降的影响，一次不能加得太多，且渣中（FeO）向钢中的扩散转移又是限制环节，而吹氧脱碳不受这种限制，因此当钢液中的碳含量（质量分数）降到 0.10% 以下时，吹氧脱碳优于加矿脱碳，且两者的速度也有显著的差别。生产实践也证明，在冶炼低碳或超低碳钢时，吹氧容易把碳很快降到很低，而且合金元素的氧化损失比矿石氧化要少，这使得利用返吹法冶炼高合金钢并回收炉料中的贵重合金元素成为可能。在其他条件相同的情况下，吹氧脱碳和加矿脱碳相比，渣中（FeO）的含量少，且钢液中 [FeO] 的最终含量也少，这样可减轻钢液精炼的脱氧负担。然而脱磷条件却恶化了，所以脱磷任务必须在熔化末期或氧化初期且当钢液的温度处于不太高的情况下就已完成。吹氧脱碳冶炼时间短，可提高产量 20% 以上，电耗降低 15%~30%，电极消耗降低 15%~30%，总成本降低 6%~8%，且钢的质量也大有改善。吹氧降碳时，最好选用较高的氧压。因为氧压高，氧气流在钢液内可吹入到更深的部位，并能分裂成更多的小气泡，从而提高氧的利用率。此外，氧压高还可减少氧管的消耗，这是由于提高了脱碳速度，缩短了吹氧时间，提高了氧的流速，强化了对管壁端的冷却作用。

3.3.2.3　Comprehensive Decarburization of Liquid Steel with Ore and Oxygen
3.3.2.3　钢液的矿、氧综合脱碳

The comprehensive decarburization of ore and oxygen increases the speed of oxygen supply to the molten pool, expands the reaction area of ore and oxygen, reduces the transfer of oxygen from steel to slag, and accelerates the diffusion speed of FeO due to the agitation of oxygen flow. Therefore, this decarburization method can make the decarburization speed of the liquid steel multiple higher than the decarburization speed of adding ore or blowing oxygen alone. In the process of operation, the ore is added in batches, and more ore is added first, then less ore is added, and finally all oxygen is used. After oxygen blowing stops, clean and boil or keep manganese.

矿、氧综合脱碳加大了向熔池供氧的速度，扩大了矿氧反应区，同时也减少了钢中氧向渣中的转移，又由于氧气流的搅动作用，使 FeO 的扩散速度加快，所以这种脱碳方法能使钢液的脱碳速度成倍地高于单独加矿或吹氧的脱碳速度。在操作过程中，矿石的加入是分批进行，且先多后少，最后全用氧气。吹氧停止后，再进行清洁沸腾或保持锰等操作。

3.3.2.4　Experience Judgment of Carbon Content
3.3.2.4　碳含量的经验判断

The carbon content of molten steel is mainly determined by chemical analysis, spectral analysis and other instruments. But in practical operation, in order to shorten smelting time, EAF steelmakers often use experience to make accurate judgment. The methods are introduced as follows:

钢液的碳含量主要依靠化学分析、光谱分析及其他仪器来确定。但在实际操作中,为了缩短冶炼时间,电炉炼钢工也常用经验进行准确的判断,方法介绍如下:

(1) The carbon content in steel is estimated according to the oxygen parameter. In the smelting process, according to the oxygen blowing time, oxygen blowing pressure, oxygen pipe insertion depth, oxygen consumption or ore addition amount, molten steel temperature, total melting carbon content, etc., the decarburization amount in 1min or a period of time is estimated first, and then the carbon content in the steel is estimated. Because this method is easy to use, it is often used.

(1) 根据用氧参数来估计钢中的碳含量。在冶炼过程中,依据吹氧时间、吹氧压力、氧管插入深度、耗氧量或矿石的加入量、钢液温度、全熔碳含量等,先估算1min或一段时间内的脱碳量,然后再估计钢中的碳含量。因这个办法使用方便,所以应用的时候较多。

(2) The carbon content in steel is estimated according to the amount of yellow smoke emitted from the furnace during oxygen blowing. The yellow smoke emitted from the furnace is thick and much, indicating that the carbon content is high, and vice versa. When the carbon content is less than 0.30%, the yellow smoke is quite light. This method can only estimate the carbon content in steel roughly, and it is difficult to make an accurate judgment.

(2) 根据吹氧时炉内冒出的黄烟多少来估计钢中的碳含量。炉内冒出的黄烟浓、多,说明碳含量高,反之较低。当碳含量(质量分数)小于0.30%时,黄烟相当淡了。这个办法只能大概地估计钢中碳含量,难以做到准确的判断。

(3) The carbon content of steel is estimated according to the spark emitted from the furnace door during oxygen blowing. When blowing oxygen, the spark ejected from the furnace door is thick and dense, with many branches, the carbon is high, and vice versa.

(3) 根据吹氧时炉门口喷出的火星估计钢中的碳含量。吹氧时炉门口喷出的火星粗密且分叉多则碳高,反之则低。

(4) The carbon content in steel is determined according to the flame condition of electrode hole during oxygen blowing. It is often used in smelting high alloy steel by back blowing method. Generally, if the carbon content is high, the flame is long; otherwise, the flame is short. When the brownish white flame contracts and there is a boiling circle between the slag and the slag line, the carbon content is generally less than 0.10%. When smelting chromium nickel stainless steel by back blowing method, when the brown white flame shrinks and purple red flame emerges and the flue gas in the furnace is small, it can be seen that the slag surface boiling is weak, and the carbon content is about 0.06% ~ 0.08%; if the slag suddenly becomes thin, it is a symbol of over blowing, and the carbon content is generally less than 0.03%. This phenomenon

is often encountered in the smelting of ultra-low carbon steel.

（4）根据吹氧时电极孔冒出的火焰状况判断钢中的碳含量，常用于返吹法冶炼高合金钢。一般是碳含量高则火焰长，反之则火焰短。当棕白色的火焰收缩，且熔渣与渣线接触部分有一沸腾圈，这时的碳含量（质量分数）一般小于 0.10%。在返吹法冶炼铬镍不锈钢时，当棕白色的火焰收缩并带有紫红色火焰冒出且炉膛中烟气不大，可见到渣面沸腾微弱，这时的碳含量（质量分数）为 0.06% ~ 0.08%；如果熔渣突然变稀，这是过吹的象征，碳含量（质量分数）一般小于 0.03%。碳低熔渣变稀，这种现象在冶炼超低碳钢时经常遇到。

(5) Make a rough judgment according to the surface tension. When the carbon content is 0.30% ~ 0.40% and less than 0.10%, the surface tension of the steel is large. When sampling, the back of the scoop slips on the steel surface.

（5）根据表面张力的大小进行粗略的判断。当碳含量（质量分数）为 0.30% ~ 0.40% 和碳含量小于 0.10% 时，钢的表面张力较大，取样时，样勺的背面在钢液面上打滑。

(6) According to the fracture characteristics of the sample, the carbon content in the steel is determined. This method is to pour the molten steel into the rectangular sample mold without deoxidation, take it out after solidification and put it into water for cooling, and then break it again. We can estimate the carbon content in the steel by using the crystal size and bubble shape of the sample fracture.

（6）根据试样断口的特征判断钢中的碳含量。这种方法是把钢液不经脱氧倒入长方形样模内，凝固后取出放入水中冷却，然后再打断，可利用试样断口的结晶大小和气泡形状来估计钢中的碳含量。

(7) The carbon content of steel is estimated according to the surface characteristics of steel cake. This method is mainly used in the smelting of low carbon steel. Generally, ladle the molten steel onto the iron plate without deoxidation, and then estimate the carbon content according to the surface characteristics of the steel cake.

（7）根据钢饼表面特征估计钢中的碳含量。这种方法主要用于低碳钢的冶炼上。一般是舀取钢液不经脱氧即轻轻倒在铁板上，然后根据形成钢饼的表面特征来估计碳含量。

(8) According to the characteristics of sparks, the carbon content in steel is identified. Using the characteristics of sparks to identify the carbon content in steel should be equipped with a grinder, but the deviation of this method is large. This is mainly because the sparks produced by grinding wheel are related to the number of revolutions of the grinder and the grain size of the grinding wheel, which is not commonly used in front of the furnace.

（8）根据火花的特征鉴别钢中的碳含量。利用火花的特征鉴别钢中的碳含量应具备砂轮机一台，但该法偏差大，这主要与砂轮打磨出的火花与砂轮机的转数及砂轮的砂粒粗细等有关，在炉前并不常用。

(9) According to the characteristics of carbon flower, the carbon content in steel is determined, which is called carbon flower observation method for short. Because the method is simple, rapid and accurate, it has been widely used.

(9) 根据碳花的特征判断钢中的碳含量,该法简称碳花观察法。由于该法简便、迅速、准确,因此获得普遍的应用。

When the molten steel without deoxidization is cooled in the scoop, it can continue to react with carbon and oxygen. When the bubble escapes, there is a very thin liquid coating on the surface, just like a hollow steel ball. This is spark. And because bubbles escape continuously, sparks often form a fire line. If there is more free carbon in steel, sometimes there are carbon particles on the surface of spark. When the pressure of the bubble is large and the strength of the bead wall is insufficient, the burst spark will break up, and then form the so-called carbon flower. However, the bubble pressure of CO decreases with the decrease of carbon content in molten steel, the number and size of carbon flowers decrease in turn, and the bursting force of spark is also from strong to weak. Experienced steelmakers can judge according to the number, size and fracture of spark (carbon flower), the strength of bursting force, the intermittent situation of fire line or the sound emitted. The lower the carbon content, the more accurate the judgment is. The error is often only ±0.01% ~0.02%. The higher the carbon content, the larger the carbon flower, the more bifurcations, the more violent the jump, and the less regularity. Therefore, the carbon content is very high, it is difficult to judge accurately in high time. When the carbon content is more than 0.80%, the carbon flower makes a squeaking sound in the process of jumping and breaking. There are two ways to observe the carbon flower: one is to directly observe the spark (carbon flower) bursting out of the spoon; the other is to observe the rupture of spark (carbon flower) after landing. See Table 3-1 for the relationship between carbon content and carbon flower characteristics of general carbon steel.

未经脱氧的钢液在样勺内冷却时能够继续进行碳氧反应,当气泡逸出时,表面附有一层很薄的液衣,宛如空心钢珠,这就是火星。又因气泡是连续逸出的,所以迸发出来的火星往往形成火线。如果钢中的游离碳较多,有时在火星的表面上还附有碳粒。当气泡的压力较大而珠壁的强度不足时,迸发出来的火星破裂,进而形成所谓的碳花。然而,CO气泡压力随钢液碳含量的降低而降低,碳花的数目和大小也依次递减,火星的迸发力量也是由强到弱。有经验的炼钢工可根据火星(碳花)的数量、大小与破裂情况及迸发力量的强弱、火线的断续情况或发出的声音等进行判断,碳含量越低判断得越准确,误差常常只有±0.01% ~0.02%,而碳含量越高,碳花越大,分叉越多,跳跃越猛烈,也越缺乏规律性,因此碳含量很高时难以准确的判断。当碳含量(质量分数)超过0.80%以上时,碳花在跳跃破裂过程中还发出吱吱的响声。碳花的具体观察方法有两种:一种是直接观察从勺内迸发出来的火星(碳花)情况;另一种是观察火星(碳花)落地后的破裂情况。一般碳钢的碳含量与碳花特征的关系见表3-1。

Table 3-1 Relationship between carbon content and carbon flower characteristics of carbon steel
表 3-1 碳钢的碳含量与碳花特征的关系

Carbon content/% $w(C)$ /%	The color 颜色	Spark and carbon flower 火星与碳花	Bursting power 迸发力量	Remarks 备注
0.05~0.10	Brownish white 棕白色	It's all made of spark 全是火星构成的火线无花	Burst out of weakness 迸发无力	Sometimes there is no fire line when it is sparse 火线稀疏时有时无
0.1~0.20	white 白色	There are little flowers in the fire line of spark 火星构成的火线中略有小花	Burst out of weakness 迸发无力	Sometimes there is no fire line when it is sparse 火线稀疏时有时无
0.3~0.40	Reddish 带红	spark 2/3, floret 1/3 火星 2/3, 小花 1/3	Burst a little bit 迸发稍有力	The fire line is thin and slightly dense 火线细而稍密
0.50~0.60	Red 红色	2/3 of hot wire, 1/3 of small flowers with 2~3 large flowers 火线 2/3, 小花 1/3, 间带 2~3 朵大花	Burst into force 迸发有力	The fire line is thick and dense 火线粗而密
0.70~0.80	Red 红色	Fire line 1/3, small flower 2/3, large flower 3~5 火线 1/3, 小花 2/3, 大花 3~5 朵	Strong burst, split inside the flower, showing a second break 迸发有力, 花内分叉, 呈现二次破裂	
0.90~1.00	Red 红色	Less spark, more small flowers, 7~10 large flowers 火星少, 小花多, 大花 7~10 朵	There are few spark, many small flowers, 7~10 big flowers bursting out forcefully and strongly, with three times of broken flowers in a circle 迸发有力、很强, 花有圈呈现三次破裂	When the carbon content is more than 0.08%, there is a squeaking noise in the process of jumping and cracking 碳含量（质量分数）大于 0.08% 以上时, 碳花在跳跃破裂过程中有吱响声
1.1~1.20	Red 红色	A lot of flowers, a little bit of spark 大花很多、很乱略有火星	The flower jumps frequently and powerfully, and there are three times of broken flowers in a circle 花跳跃频繁有力, 花有圈呈现三次破裂	

Continued Table 3-1
续表 3-1

Carbon content/% $w(C)/\%$	The color 颜色	Spark and carbon flower 火星与碳花	Bursting power 迸发力量	Remarks 备注
1.30~1.40	Red 红色	Big flower 1/3, purple flower 2/3 大花1/3，紫花2/3	Flowers jump short and strong, breaking many times 花跳跃短而有力，多次破裂	
1.50~1.80	Red 红色	Almost all purple flowers 几乎全是紫花	Flowers jump, rain is short and strong, and break many times 花跳跃短而有力，多次破裂	

In order to judge the carbon content of steel by the characteristics of carbon flower, the following factors should also be considered. If the temperature is too low, it is easy to look low, but the actual carbon content is not so low; if the temperature is too high, it is easy to look high, but the actual carbon content is not so high. The carbon flower of alloy steel is similar to that of carbon steel, but its shape is different due to the influence of other elements. Compared with carbon steel with the same carbon content: when the content of Mn, Cr, V and other elements is higher, the carbon flower is larger, more bifurcated, and easy to estimate is on the high side; when the content of W, Ni, Si, Mo and other elements is higher, the carbon flower is less bifurcated, and easy to estimate is on the low side. In addition, if the oxygen content in steel is lower than the equilibrium value of carbon and oxygen, there will be less carbon flowers or no carbon flowers, such as deoxidized or vacuum treated liquid steel.

利用碳花的特征判断钢中的碳含量，还应考虑以下一些因素的影响。温度过低，容易低看，而实际碳含量不是那么低；温度过高，容易高看，而实际碳含量又没有那么高。合金钢的碳花与碳钢基本相似，但形状因其他元素的影响而有所不同。与碳含量相同的碳钢比较：当Mn、Cr、V等元素含量较高时，碳花较大、分叉又多，容易估的偏高；当W、Ni、Si、Mo等元素含量较高时，碳花分叉较少，容易估的偏低。除此之外，如钢中的氧含量低于碳氧的平衡值时，碳花较少或无碳花，如经脱氧或真空处理的钢液就是如此。

3.3.3 Oxidation of Iron, Silicon, Manganese, Chromium and Other Elements
3.3.3 铁、硅、锰、铬等元素的氧化

At the same time of dephosphorization and decarbonization, other elements and impurities in the molten steel are oxidized. The oxidation and oxidation degree of these elements depend on the affinity of the element to oxygen compared with iron, the concentration of the element in the molten steel, the composition of the slag and the chemical properties (acid or alkaline) and smelting temperature of the element oxide. In addition, in the actual smelting, the oxidation degree of each element in the molten steel is also affected by the oxidation reaction speed of each element, which is related to the physical properties of the slag (viscosity, surface tension) and the physical state of the molten pool.

在脱磷、脱碳反应进行的同时，钢液中其他元素和杂质也发生氧化。这些元素的氧化

及氧化程度取决于：与铁比较该元素对氧的亲和力大小、该元素在钢液中的浓度、熔渣组成与该元素氧化物的化学性质（酸性或碱性）及冶炼温度等。除此之外，在实际冶炼中，钢液中各元素的氧化程度还受各元素氧化反应速度的影响，而这速度又与熔渣的物理性质（黏度、表面张力）及熔池的物理状态有关。

Aluminum has a strong affinity for oxygen, followed by titanium. After melting and oxidation of molten steel, almost all of them are oxidized, while Fe, Si, Mn, Cr and other elements are different from aluminum and titanium. Let's introduce them respectively.

铝对氧的亲和力很强，钛次之，钢液经熔化和氧化后，它们几乎全部氧化掉，而Fe、Si、Mn、Cr等元素与铝和钛的情况不一样。下面分别进行介绍。

3.3.3.1　Oxidation of Iron
3.3.3.1　铁的氧化

The affinity between iron and oxygen is smaller than that of other elements (except Cu, Ni, CO). As long as there are other elements, it is difficult to combine with oxygen. However, in iron-based molten steel, because the concentration of iron is the largest, it is still oxidized first. When C, Si, Mn, P and other elements in steel are oxidized, the main source of oxygen is FeO, so the oxidation of iron can store oxygen for the oxidation of other elements, which also shows that FeO plays an important role in oxygen transfer.

铁与氧的亲和力比其他元素（Cu、Ni、CO除外）小，只要有别的元素存在，它就很难与氧结合，但在铁基钢液中，由于铁的浓度最大，所以它还是首先被氧化。当钢中的C、Si、Mn、P等元素进行氧化反应时，氧的主要来源是FeO，因此铁的氧化能为其他元素的氧化贮存氧，这也说明FeO对氧的传递起着重要作用。

3.3.3.2　Oxidation of Silicon
3.3.3.2　硅的氧化

Si has a higher affinity with O, but is inferior to Al and Ti, but stronger than V, Mn and Cr. Therefore, Si in molten steel will be oxidized by 70% in the melting period, and a small amount of residual Si will be reduced to the minimum in the early stage of oxidation. The oxidation reaction formula of silicon is as follows:

$$[Si] + 2(FeO) = (SiO_2) + 2[Fe]$$
$$[Si] + O_2 = SiO_{2(s)}$$

Si与O的亲和力较大，但次于Al和Ti，而强于V、Mn、Cr。因此，钢液中的Si在熔化期将被氧化掉70%，少量的残余Si在氧化初期也能降低到最低限度。硅的氧化反应式如下：

$$[Si] + 2(FeO) = (SiO_2) + 2[Fe]$$
$$[Si] + O_2 = SiO_{2(固)}$$

The oxidation of silicon is exothermic reaction, and the degree of oxidation decreases with the increase of temperature, and the oxidation is complete in alkaline slag than in acid slag. When smelting high chromium stainless steel or back blowing other steel grades, a certain amount of silicon is often added into the furnace charge to ensure the rapid temperature rise of the molten steel and reduce the burning loss of chromium. SiO_2, the reaction product, is insoluble in the molten steel. In addition to part of it can float up to the slag, some of it can be suspended in the molten steel as fine particles. SiO_2 can combine with other inclusions to form silicate. If it is removed improperly, it can become inclusions in the steel.

硅的氧化是放热反应,并随着温度的升高氧化程度减弱,且在碱性渣下比在酸性渣下氧化完全。在冶炼高铬不锈钢或返吹其他钢种时,炉料中往往配入一定量的硅,以保证钢液的快速升温及减少铬的烧损。反应生成物 SiO_2 不溶于钢液中,除一部分能上浮到渣中外,还有一部分呈细小颗粒夹杂悬浮于钢液中,SiO_2 可与其他夹杂物结合生成硅酸盐,如果去除不当而残留在钢中将成为夹杂物。

In the slag, (SiO_2) reacts with (FeO) as follows:

$$(SiO_2) + 2(FeO) = (2FeO \cdot SiO_2)$$

In the alkaline slag, (FeO) can be replaced by stronger alkaline oxide (CaO) from ferric silicate, namely:

$$(2FeO \cdot SiO_2) + 2(CaO) = (2CaO \cdot SiO_2) + 2(FeO)$$

As the generated calcium orthosilicate ($2CaO \cdot SiO_2$) is a stable compound, the decomposition pressure of SiO_2 becomes lower, and the $a_{(SiO_2)}$ in the slag becomes smaller, so the oxidation of silicon is complete in the alkaline slag.

在熔渣中,(SiO_2) 与 (FeO) 发生下述反应:

$$(SiO_2) + 2(FeO) = (2FeO \cdot SiO_2)$$

在碱性渣中,(FeO) 能被更强的碱性氧化物 (CaO) 从硅酸铁中置换出来,即:

$$(2FeO \cdot SiO_2) + 2(CaO) = (2CaO \cdot SiO_2) + 2(FeO)$$

由于生成的正硅酸钙 ($2CaO \cdot SiO_2$) 是稳定的化合物,使 SiO_2 分解压力变得更低,渣中 $a_{(SiO_2)}$ 变得更小,所以在碱性渣下,硅的氧化较完全。

3.3.3.3 Oxidation of Manganese
3.3.3.3 锰的氧化

The affinity between manganese and oxygen is smaller than that of silicon, and the loss of manganese is about 50% at the end of melting. If the FeO content and basicity are low, the burning loss may be less.

锰与氧的亲和力比硅小,到了熔化末期锰大约烧损 50%。如果熔化期渣中 (FeO) 含量和碱度较低时,烧损可能还要少些。

During the oxidation period, manganese in the molten steel continues to oxidize, and the oxidation reaction of manganese is as follows:

$$[Mn] + (FeO) = (MnO) + [Fe]$$

在氧化期,钢液中的锰继续氧化,锰的氧化反应如下:

$$[Mn] + (FeO) = (MnO) + [Fe]$$

The oxidation of manganese is also exothermic reaction, and the degree of oxidation decreases with the increase of temperature. When the temperature of molten steel increases to a certain extent, the oxidation of manganese tends to balance. Therefore, when the content of manganese in molten steel is high after full melting, it can be oxidized at a lower temperature in the early stage of oxidation, and it can be removed by means of automatic slag flow or slag exchange.

锰的氧化也是放热反应，且随着温度的升高氧化程度减弱。当钢液的温度升高到一定程度时，锰的氧化反应趋于平衡。因此，全熔后钢液中的锰含量较高时，可在氧化初期在较低的温度下进行氧化，并采取自动流渣或换渣的方式去除。

Usually in the actual smelting process of EAF steel, the change of manganese content is regarded as a sign of the temperature of molten steel. This is because after the temperature of the molten pool increases, the content of FeO in the slag decreases continuously due to the oxidation reaction of carbon, and the free MnO dissolved in the slag will participate in the oxidation reaction of carbon. At this time, the oxidation reaction of manganese which has been in equilibrium will be destroyed and converted into the reduction of manganese. The reaction formula is as follows:

$$(MnO) + [Fe] = [Mn] + (FeO)$$
$$\underline{(FeO) + [C] = [Fe] + \{CO\}}$$
$$(MnO) + [C] = [Mn] + \{CO\}$$

通常在电炉钢的实际冶炼过程中，锰含量的变化被看成是钢液温度高低的标志。这是因为熔池温度升高后，由于碳的氧化反应使渣中的（FeO）含量不断降低，溶于熔渣中呈游离状态的（MnO）就要参与碳的氧化反应，这时已趋于平衡的锰的氧化反应被破坏，而转变为锰的还原，反应式如下：

$$(MnO) + [Fe] = [Mn] + (FeO)$$
$$\underline{(FeO) + [C] = [Fe] + \{CO\}}$$
$$(MnO) + [C] = [Mn] + \{CO\}$$

If the manganese content in the molten steel at the end of oxidation is higher than that at the beginning or there is no loss of manganese in the oxidation process, the oxidation boiling is carried out at high temperature. If the loss of manganese content in molten steel is more at the end of oxidation, it indicates that the oxidation boiling may be carried out at a lower temperature.

如果氧化末期钢液中的锰含量比前期高或氧化过程中锰元素没有损失，说明氧化沸腾是在高温下进行的。如果氧化末期钢液中的锰含量损失较多，说明氧化沸腾有可能是在较低的温度下进行的。

The solubility of manganese oxide (MnO) in the molten steel is very small, and it will float up into the slag. Part of MnO is combined with SiO_2 and Al_2O_3, which are suspended in the molten steel, to form manganese silicate and manganese aluminate:

$$m(MnO) + n(SiO_2) = (mMnO \cdot nSiO_2)$$
$$m(MnO) + n(Al_2O_3) = (mMnO \cdot nAl_2O_3)$$

锰的氧化反应生成物（MnO）在钢液中的溶解度很小，将上浮进入渣中，其中一部分在上浮途中与悬浮在钢液中的细小不易上浮的 SiO_2、Al_2O_3 结合成硅酸锰和铝酸锰即：

$$m(MnO) + n(SiO_2) = (mMnO \cdot nSiO_2)$$

$$m(MnO) + n(Al_2O_3) = (mMnO \cdot nAl_2O_3)$$

These manganous silicates and manganous aluminates are compounds with large particle size, low melting point and easy to float. In order to remove the inclusions in the steel better, the process system of keeping manganese was established at the end of oxidation when smelting the steel with important use or low carbon content. When the slag has enough basicity, (CaO) can replace (MnO) from manganese silicate or manganese aluminate to form a relatively stable and difficult to decompose compound, namely:

$$(2MnO \cdot SiO_2) + 2(CaO) = (2CaO \cdot SiO_2) + 2(MnO)$$
$$(MnO \cdot Al_2O_3) + (CaO) = (CaO \cdot Al_2O_3) + (MnO)$$

这些硅酸锰和铝酸锰属于颗粒大、熔点低并易于上浮的化合物。为了更好地去除钢中夹杂，在冶炼用途重要或碳含量较低的钢种时，在氧化末期建立了保持锰的工艺制度。当熔渣具有足够的碱度时，（CaO）能将（MnO）从硅酸锰或铝酸锰中置换出来而形成比较稳定不易分解的复合化合物，即：

$$(2MnO \cdot SiO_2) + 2(CaO) = (2CaO \cdot SiO_2) + 2(MnO)$$
$$(MnO \cdot Al_2O_3) + (CaO) = (CaO \cdot Al_2O_3) + (MnO)$$

3.3.3.4　Oxidation of Chromium
3.3.3.4　铬的氧化

Cr is easier to oxidize than Fe, but less than Al, Ti, Si, etc. the oxidation reaction formula of Cr is as follows:

$$[Cr] + [O] = (CrO)$$
$$2[Cr] + 3(FeO) = (Cr_2O_3) + 3[Fe]$$

Cr 比 Fe 易氧化，但不如 Al、Ti、Si 等，铬的氧化反应式如下：

$$[Cr] + [O] = (CrO)$$
$$2[Cr] + 3(FeO) = (Cr_2O_3) + 3[Fe]$$

The second reaction is the main one for oxidation under alkaline slag. The reaction formula also shows that the recovery rate of chromium can be reduced by FeO, and when the chromium content in the charge is very high, the chromium loss transferred into the slag is also large. The slag with high chromium content is very sticky, which can affect the normal oxidation of other elements (such as dephosphorization). Therefore, in order to reduce the loss of chromium and ensure the normal operation of smelting, it is not suitable to use high chromium burden in ore oxidation.

在碱性渣下的氧化，第二个反应是主要的。该反应式也表明了渣中（FeO）能使铬的收得率降低，且当炉料中的铬含量很高时，转入熔渣中的铬损失也多。高铬的熔渣很黏，能影响其他元素氧化反应（如脱磷）的正常进行。因此，为了减少铬的损失及保证冶炼的正常进行，矿石法氧化不宜使用含高铬的炉料。

The oxidation of chromium can also release a lot of heat, and the oxidation loss of chromium is related to temperature. In addition, the following reactions will occur at high temperature:

$$(CrO) + [C] = [Cr] + \{CO\}$$

$$(Cr_2O_3) + 3[C] = 2[Cr] + 3\{CO\}$$

铬的氧化也能放出大量的热，而铬的氧化损失又与温度有关。除此之外，在高温时还将发生下述反应：

$$(CrO) + [C] = [Cr] + \{CO\}$$
$$(Cr_2O_3) + 3[C] = 2[Cr] + 3\{CO\}$$

Decarburization at high temperature can inhibit the oxidation loss of chromium, which is of great significance for smelting low carbon and high chromium steel by back blowing method. In other words, high temperature is not conducive to the oxidation of chromium, so in order to reduce the chromium content in steel, generally low temperature and ore oxidation are used.

高温下脱碳能抑制铬的氧化损失，这一点对于采用返吹法冶炼低碳高铬钢具有极其特殊的意义。换言之，高温不利于铬的氧化，所以为了降低钢中的铬含量，一般均采用偏低的温度并选用矿石氧化的方式进行。

3.3.3.5　Oxidation of Vanadium
3.3.3.5　钒的氧化

Vanadium has a high affinity for oxygen. When the content of FeO is very high, it almost oxidizes. The oxidation of vanadium is also exothermic reaction, so lower oxidation temperature can increase the oxidation loss of vanadium. The reaction formula is as follows:

$$2[V] + 3(FeO) = (V_2O_3) + 3[Fe]$$
$$2[V] + 5(FeO) = (V_2O_5) + 5[Fe]$$

钒对氧的亲和力较大，当熔池中（FeO）的含量很高时，它几乎全部氧化。钒的氧化也是放热反应，因此偏低的氧化温度可使钒的氧化损失增加。反应式如下：

$$2[V] + 3(FeO) = (V_2O_3) + 3[Fe]$$
$$2[V] + 5(FeO) = (V_2O_5) + 5[Fe]$$

With the increase of temperature, a large amount of (FeO) is captured due to the intense oxidation of carbon, which can also inhibit the oxidation of vanadium. Vanadium is easy to oxidize and the products (V_2O_3 and V_2O_5) are also easy to reduce. If there are more vanadium bearing slag left in the furnace, or there are more vanadium bearing slag hanging on the furnace wall and furnace cover, a part of vanadium will be reduced back to the steel in the reducing atmosphere of electric furnace steelmaking. In this case, special attention should be paid when smelting vanadium steel with vanadium bearing charge to avoid the composition of vanadium exceeding the specification.

温度升高后，由于碳的激烈氧化夺取了大量的（FeO），从而也能抑制钒的氧化。钒既易氧化，产物（V_2O_3）、（V_2O_5）又极易还原。如果炉中剩留的含钒氧化渣较多，或炉壁和炉盖处悬挂的含钒氧化渣较多，在电炉炼钢的还原气氛下，将有一部分的钒被还原回钢中。这种情况在利用含钒炉料冶炼钒钢时尤要注意，以避免钒的成分超出规格。

3.3.3.6 Oxidation of Tungsten
3.3.3.6 钨的氧化

Tungsten is a weak reducing agent, which is easier to oxidize than iron. In the oxidation process of EAF steel, when the content of (FeO) is very high, the oxidation and burning loss of tungsten is also very serious. The reaction formula is as follows:

$$[W] + 3(FeO) = (WO_3) + 3[Fe]$$
$$2[W] + 5(FeO) = (W_2O_5) + 5[Fe]$$
$$[W] + 2(FeO) = (WO_2) + 2[Fe]$$

钨是一种弱还原剂,它比铁容易氧化。在电炉钢的氧化过程中,当(FeO)的含量很高时,钨的氧化烧损也很严重。反应式如下:

$$[W] + 3(FeO) = (WO_3) + 3[Fe]$$
$$2[W] + 5(FeO) = (W_2O_5) + 5[Fe]$$
$$[W] + 2(FeO) = (WO_2) + 2[Fe]$$

Among them, WO_3 is an acid oxide, in addition to part of which can be reduced, when the alkalinity in slag reaches enough high, the following reactions can occur:

$$(WO_3) + (CaO) = (CaWO_4)$$

其中,WO_3 为酸性氧化物,除有一部分能被还原外,当渣中碱度达到足够高时,又能发生下述反应:

$$(WO_3) + (CaO) = (CaWO_4)$$

Therefore, relatively speaking, the acid smelting of tungsten is almost free of loss, while the alkaline smelting has a large loss. In addition, the boiling point of (WO_3) is about 1850℃, so part of (WO_3) may sublimate under the light column of the arc to volatilize tungsten. It is not difficult to find out that the ore oxidation can cause the loss of tungsten in the molten steel, and the loss of alkaline smelting is greater. Because the production of ferrotungsten is difficult and the price is high, the burden containing tungsten should not be used in ore smelting. However, tungsten has high melting point, high density and is easy to deposit at the bottom of furnace. If the content of [FeO] is not high, the oxidation loss of tungsten is also limited, which makes it possible for us to smelt high tungsten steel by loading method or back blowing method.

因此,相对比较而言,酸性熔炼钨几乎不受损失,而碱性熔炼损失较大。此外 (WO_3) 的沸点约为1850℃,所以还有一部分的 (WO_3) 可能在电弧的光柱下升华而使钨挥发。不难得出,矿石法氧化能使钢液中的钨蒙受损失,且碱性熔炼损失更大。因钨铁的生产比较困难,价格较高,所以矿石法冶炼不应使用含钨的炉料。然而钨的熔点高,密度大,易沉积炉底,如果[FeO]的含量不高时,钨的氧化损失也是有限的,这样就为利用装入法或返吹法冶炼高钨钢提供了可能。

3.3.3.7 Oxidation of Molybdenum, Nickel, Cobalt, Copper and Other Elements
3.3.3.7 钼、镍、钴、铜等元素的氧化

The affinity of molybdenum to oxygen is almost the same as that of iron. In the smelting process of electric furnace steel, if the content is not high, its oxidation loss is very small, which can be ignored generally. However, when smelting high molybdenum steel ($w(Mo)>4\%$), the oxidation loss must be considered, because the oxidation and oxidation degree of molybdenum increase with the increase of molybdenum content in steel, that is to say, the oxidation loss of molybdenum is related to the concentration of molybdenum in molten steel. The affinity of Ni, Co and Cu to oxygen is much smaller than that of iron, and they will not be oxidized under the condition of steelmaking, but sometimes Ni also has loss, which is the result of volatilization in the high temperature zone of arc. Through oxidation, as, Sb and Sn are difficult to be removed in the oxidation stage.

钼对氧的亲和力几乎与铁一样，在电炉钢的冶炼过程中，如含量不高，它的氧化损失很微小，一般可忽略不计。但在冶炼高钼钢（$w(Mo)>4\%$以上）时，氧化损失必须予以考虑，这是因为钼的氧化与氧化程度随钢中钼含量的增加而增加，即钼的氧化损失与钼在钢液中的浓度有关。Ni、Co、Cu对氧的亲和力比铁小很多，在炼钢条件下不会被氧化，但Ni有时也有损失，那是在电弧高温区挥发的结果。通过氧化方式，As、Sb、Sn元素在氧化期一般是很难去除的。

3.3.4 Establishment of Smelting Temperature System and Temperature Rise of Molten Steel
3.3.4 冶炼温度制度的制订和钢液的升温

EAF steelmaking pays great attention to smelting temperature, because it regulates and affects the direction and limit of reaction, and directly relates to the quality, output and various technical and economic indexes of steel. Therefore, it has always been a major task for the operation of EAF steelmakers to master the smelting temperature correctly.

电炉炼钢十分讲究冶炼温度，因为它规定和影响着反应的方向与限度，直接关系到钢的质量、产量及各项技术经济指标。因此，正确掌握冶炼温度历来都是电炉炼钢工操作的一项主要任务。

3.3.4.1 Establishment of Smelting Temperature System
3.3.4.1 冶炼温度制度的制订

Because temperature has a great influence on EAF steelmaking, it is an important parameter in smelting process. In order to ensure the smooth operation of smelting, it is necessary to establish a suitable smelting temperature system according to the specific situation. The following factors should be taken into account in the formulation of temperature system:

由于温度对电炉炼钢的影响很大，因此它是冶炼工艺中的一个重要参数。为了确保冶炼的顺利进行，针对具体情况制订合适的冶炼温度制度是十分必要的。温度制度的制订主要应考虑以下诸因素：

(1) Master the melting point of steel and the viscosity of liquid steel. A large number of scientific tests and production practices have proved that only when the smelting temperature exceeds a

certain range of melting point, can various physical and chemical reactions between slag steel be fully carried out, so the melting point of steel is the basis for formulating smelting temperature system. The composition of molten steel is constantly changing in the smelting process, so when making the smelting temperature system, it is necessary to calculate the melting point approximately according to the composition of molten steel in different periods and relevant empirical formulas. For example, the melting points of 18CrMnTi and 12CrNi3A are 1510℃, but when the temperature is increased by 100℃, the viscosity of 12CrNi3A changes little, while that of 18CrMnTi decreases by 33%. Because the viscosity of molten steel is different at the same temperature, the diffusion and mass transfer of elements and the speed of various physicochemical reactions are also different. Therefore, in order to ensure the smooth progress of metallurgical process, the influence of the viscosity of molten steel must be taken into account when formulating the smelting temperature system.

（1）掌握钢的熔点与钢液的黏度。大量的科学试验和生产实践证明，只有冶炼温度超过熔点的一定范围时，渣钢间的各种物化反应才能得以充分进行，因此钢的熔点是制订冶炼温度制度的基础。而冶炼过程中钢液的成分是不断变化的，所以在制订冶炼温度制度时，必须依据不同时期的钢液成分，并运用有关的经验式对熔点进行近似计算。不同的钢种在相同的温度下黏度相差较大，如18CrMnTi和12CrNi3A的熔点均为1510℃，但当温度升高100℃时，12CrNi3A的黏度变化不大，而18CrMnTi的黏度却降低33%。由于在相同的温度下，钢液的黏度不同，元素的扩散与传质及各种物化反应速度等也不同，因此为了保证冶金过程的顺利进行，在制订冶炼温度制度时就必须考虑钢液黏度的影响。

（2）Familiar with the temperature range of various physicochemical reactions. In addition to mastering the melting point of steel and the viscosity of molten steel, the formulation of smelting temperature system is to be familiar with the temperature range of various physicochemical reactions. For example, dephosphorization and decarbonization in the oxidation stage: dephosphorization is mainly in the early stage, and the temperature should be moderately low; decarbonization is mainly in the later stage, and the temperature should gradually increase. In addition, the melting point of steel gradually rises due to the oxidation of carbon, and the decarburization amount in the oxidation period is generally greater than 0.30%. For the liquid steel with carbon content less than 1.0%, the melting point is correspondingly increased by more than 20℃, so the temperature in the later stage of oxidation should be higher. Deoxidization and desulfuration are the main reactions in the reduction period of EAF steelmaking. Deoxidization is the key. Considering the characteristics of deoxidization process and the original sulfur content in steel, smelting should be carried out at a higher temperature. However, after the reduction of molten steel, due to the continuous introduction of a large number of deoxidizers and alloy elements, the melting point of steel is reduced, so the smelting temperature in the reduction period should be gradually reduced from high. For example, ZGMn13 has low melting point and good fluidity, but when smelting temperature is above 1600℃, manganese in steel will react with SiO_2 in refractories, which is easy to cause serious damage to furnace lining or furnace leakage accident. Therefore, the temperature range of this special reaction should not be ignored in the formulation of ZGMn13 steel smelting temperature system.

（2）熟知各种物化反应的温度范围。冶炼温度制度的制订除了要掌握钢的熔点和钢液的黏度外，就是要熟知各种物化反应的温度范围。例如，氧化期的脱磷和脱碳：前期主要是脱磷，温度应中等偏低；后期主要是脱碳，温度应逐渐升高。另外，钢的熔点因碳的氧化而逐渐上升，氧化期的脱碳量（质量分数）一般大于 0.30%，对于碳含量（质量分数）小于 1.0% 的钢液，熔点相应提高 20℃ 以上，因此氧化后期的温度应更高些。电炉炼钢的还原期主要进行脱氧、脱硫反应，其中脱氧是关键，考虑脱氧工艺特点及钢中原始硫含量，冶炼应在较高的温度下进行。但是，钢液还原后，由于大量脱氧剂和合金元素的不断引入，钢的熔点是降低的，因而还原期的冶炼温度又理应是由高逐渐降低的。如 ZGMn13 的熔点低，流动性较好，但当冶炼温度在 1600℃ 以上时，钢中的锰将与耐火材料中的 SiO_2 发生反应，易使炉衬损坏严重或造成漏炉事故，所以在制定 ZGMn13 钢冶炼温度制度时，不可不考虑这一特殊反应的温度范围。

(3) Understand the characteristics of steel grades and metallurgical defects. Different kinds of steel have different characteristics, which lead to different metallurgical defects and different requirements for smelting temperature. Generally, for steel grades with low carbon content, high melting point and high viscosity, the smelting temperature should be slightly higher, while for steel grades with high carbon or silicon or manganese content and good fluidity, it should be slightly lower. For the steel grades sensitive to white spot, segregation and lamellar fracture, the smelting temperature should be lower, while for the steel grades requiring inspection of hairline, it should be higher.

（3）了解钢种的特性和易产生的冶金缺陷。不同的钢种有不同的特性，易产生不同的冶金缺陷，对冶炼温度的要求也不一样。一般对于碳含量低、熔点高、黏度大的钢种，确定的冶炼温度要略高一些，而对于含碳、含硅或含锰高，流动性好的钢种应稍低些。对于白点、偏析、层状断口敏感的钢种，确定的冶炼温度要适当低一些，而对于要求检查发纹的钢种要适当高一些。

(4) Various heat generation and temperature drop in smelting process are considered. All kinds of heat generated in the smelting process should be considered in the formulation of smelting temperature system. For example, the temperature of the molten pool is increased by blowing oxygen in the melting period or oxidation of C, Si, Mn and other elements; the reaction of many deoxidizing elements with oxygen in the reduction period is also exothermic reaction, when the amount is large, if not considered, high temperature steel is easy to appear and waste a lot of heat. In addition, when formulating smelting temperature system, various temperature drops in smelting process should also be considered, such as the temperature drop in the final stage of oxidation, the temperature drop in slag making materials and various ferroalloys from normal temperature to melting point, then from solid state to liquid endothermic temperature drop, the temperature drop in intermediate tapping, tapping process and ladle refining or using solid synthetic slag, as well as the temperature drop in sedation and pouring process If neglected, it is easy to cause smelting of low-temperature steel or post heating.

（4）考虑冶炼过程中的各种生成热和温度降。冶炼温度制度的制订应考虑冶炼过程中的各种生成热。如熔化期的吹氧助熔或 C、Si、Mn 等元素的氧化，均使熔池的温度升高；还原期许多脱氧元素与氧发生的反应也是放热反应，当用量较多时，如果不考虑，容易出现高温钢并浪费大量的热量。此外，在制订冶炼温度制度时，还应考虑冶炼过程中的各种温度降，如氧化末期的全扒渣降温、造渣材料和各种铁合金从常温加热到熔点，再由固态转变为液态的吸热降温、中间出钢法的降温、出钢过程及包中精炼或采用固体合成渣的降温，镇静与浇注过程的降温等，如果忽视，易使冶炼出现低温钢或后升温。

3.3.4.2　Temperature Rise of Molten Steel
3.3.4.2　钢液的升温

Due to dephosphorization, from the end of melting to the early stage of oxidation, the temperature of molten steel is mostly medium low. However, in order to ensure the intense boiling of the molten pool and the decarburization reaction at high temperature, the melting of the ore also needs to consume a certain amount of heat. With the decrease of carbon content and the increase of the melting point of the steel, it is required that the molten steel in the oxidation period must be heated continuously. In addition, the temperature rise of molten steel during the oxidation period also creates favorable conditions for slagging, deoxidization, desulfurization and alloying during the reduction period.

由于脱磷，熔化末期至氧化前期，钢液的温度多是中等偏低的。但为了保证熔池的激烈沸腾，脱碳反应在高温下进行，矿石的熔化也需要消耗一定的热量，随着碳含量的降低而钢的熔点提高，要求氧化期的钢液必须不断地进行升温。此外，氧化期钢液的升温也为还原期造渣、脱氧、脱硫及合金化等创造了有利条件。

From the analysis of the heating conditions of the melt, the temperature rise of the molten steel in the oxidation period is better than that in the reduction period. This is because the boiling caused by the intense carbon oxygen reaction accelerates the heat transfer, which can rapidly increase the temperature of the molten steel in a short time and make the temperature of the molten steel more uniform. However, in the reduction stage, the molten pool is relatively calm, and the viscosity of molten steel increases due to the introduction of alloy elements and deoxidizers, which is not conducive to the thermal vibration of molecules or the passage of free electrons. In addition, the thermal conductivity of reducing steel liquid and oxidizing steel liquid and reducing slag and oxidizing slag are different. The total content of alloying elements and impurities in the reducing steel is higher than that in the oxidizing steel. Under the same conditions, the thermal conductivity of the reducing steel is lower than that of the oxidizing steel. In the same way, the thermal conductivity of reducing slag is not as good as that of oxidizing slag. Generally speaking, the thermal conductivity of static slag is $20 \sim 40$ times lower than that of boiling slag. The arc heat of EAF steelmaking is transmitted to the molten steel through the slag. Although a large number of alloy elements and deoxidizers are introduced in the reduction period, the melting point of the steel has decreased, but the general situation is still not conducive to the temperature rise and heating of the molten steel.

从熔体的加热条件上分析，氧化期钢液的升温比还原期有利，这是因为激烈的碳氧反应引起的沸腾加速了热量的传递，可在短时间内迅速提高钢液温度以及使钢液温度变得更加均匀。而还原期的熔池比较平静，钢液中因合金元素与脱氧剂的引入，黏度增加，不利于分子的热振动或自由电子的穿过。另外，还原性钢液和氧化性钢液及还原渣和氧化渣的热导率不一样。在还原性钢液中，合金元素和杂质的总含量比氧化性钢液中的高，在其他条件相同的情况下，还原性钢液的热导率不如氧化性钢液的高。同理，还原渣的导热能力也不如氧化渣好。一般说来，静止熔渣的热导率要比沸腾熔渣的热导率低 20~40 倍。电炉炼钢的弧光热是通过熔渣传给钢液的，尽管还原期引入了大量的合金元素与脱氧剂，钢的熔点虽然下降了，但总的情况仍然不利于钢液的升温与加热。

The temperature rise of liquid steel in the later stage of reduction is commonly known as post heating. Due to the poor thermal conductivity of the melt, the speed of post heating is very slow and the temperature distribution in the molten pool is uneven, which not only prolongs the reduction time, but also often results in poor deoxidization effect, thus increasing the gas and inclusion content in the steel, and reducing the service life of the furnace lining. It is not conducive to operation, but also affects the quality of steel. Therefore, in the smelting process, the temperature of molten steel at the end of oxidation should be higher than or at least equal to the tapping temperature.

钢液在还原后期的升温俗称后升温。由于熔体的导热性能不好，后升温的速度极为缓慢且熔池中温度分布也不均匀，既延长了还原时间，又往往导致较差的脱氧效果，进而使钢中气体和夹杂物含量增加，也降低炉衬的使用寿命。既不利于操作，又影响钢的质量。因此，在冶炼过程中要求氧化末期钢液的温度要高于或至少应等于出钢温度。

However, the temperature rise of molten steel cannot be too high endlessly, because the high smelting temperature not only wastes a lot of raw materials and electric energy, but also erodes the furnace lining seriously. For the large capacity furnace platform, because of the high enthalpy and the difficulty of cooling, it is easy to produce high temperature steel or affect the reduction operation and steel pouring.

然而，钢液的升温又不能无止境的过高，因过高的冶炼温度不仅浪费大量的原材料与电能，而且严重侵蚀炉衬。对于容量大的电炉炉台，由于热焓高，降温困难，易出现高温钢或影响还原操作及钢的浇注。

3.3.5　Full Slagging and Carburization
3.3.5　全扒渣与增碳

3.3.5.1　Full Slagging
3.3.5.1　全扒渣

Full slagging is to remove all the molten slag. After the oxidation, the molten pool will turn into the reduction period, but in order to overcome the oxidation state in the furnace and prevent the reduction of harmful impurities in the slag, it is necessary to remove all the oxide slag. Total slagging is the boundary between oxidation and reduction. In order to dephosphorize or remove chromium at the end of melting, to desulfurate or cool down before tapping, or to make the recovery rate of some easily oxidized elements high and stable, sometimes it is also necessary to carry out full slagging. Here we refer to the full slagging at the end of oxidation.

全扒渣就是将熔融炉渣全部扒除。氧化结束后，熔池即将转入还原期，但为了迅速克服炉内的氧化状态以及防止熔渣中有害杂质的还原，需将氧化渣全部扒除。全扒渣是氧化与还原的分界线。熔化末期为了脱磷或去铬，出钢前为了脱硫或降温或使某些易氧化元素收得率高且又稳定，有时也需要进行全扒渣，本书指的是氧化末期的全扒渣。

（1）Conditions of fully slagging at the end of oxidation. After the completion of various tasks in the oxidation period, the molten steel needs to remain in the furnace for refining. At this time, the full slagging operation can only be carried out if the following conditions are met：

（1）氧化末期的全扒渣条件。氧化期各项任务完成后，钢液需要继续留在炉内进行精炼，这时只有具备下述条件方可进行全扒渣操作：

1）High enough temperature for slagging. At the end of oxidation, the temperature of molten steel is generally 20~30℃ higher than the tapping temperature, which should be at least equal to the tapping temperature (the exception is that a large number of ferroalloy is added under the thin slag). This is mainly considered that a large number of slagging materials, ferroalloys and slagging process are will reduce the temperature of molten steel. In addition, in order to form the thin slag as soon as possible, prevent the temperature rise after the reduction period and ensure the smooth progress of deoxidization, desulfurization and other metallurgical reactions, it is also required that the molten steel should have a high temperature for slag raking.

1）足够高的扒渣温度。氧化末期钢液温度一般均高于出钢温度20~30℃，至少也应等于出钢温度（稀薄渣下加入大量铁合金的例外），这主要是考虑到扒渣时和扒渣后加入大量的造渣材料与铁合金以及扒渣过程等都会降低钢液温度。此外，为了尽快形成稀薄渣和防止还原期后升温及保证脱氧、脱硫等冶金反应的顺利进行，也要求钢液要有足够高的扒渣温度。

2）Suitable chemical composition. The carbon content in the steel shall reach the required range: general carbon steel shall reach the lower limit of specification; carbon tool steel shall reach the lower limit of specification; the carbon content of alloy steel slags shall be added with the carbon content brought in by ferroalloy, plus the carbon content of slagging materials and deoxidization process to reach the lower limit of specification. The phosphorus content can only increase but not decrease in the reduction period, which is caused by the incomplete slag removal or the reduction of phosphorus in the slag suspended on the furnace wall and furnace cover. In addition, the phosphorus added to the ferroalloy is also brought into the steel. Therefore, the lower the phosphorus content in the molten steel, the better. Generally, the phosphorus content of various specifications shall meet the requirements in Table 3-2 before slagging. For manganese content, when smelting steel with low manganese content (such as steel T10, etc.), the manganese content before fully slagging shall be controlled according to the requirements of Table 3-3. For high-grade structural steel or liquid steel with carbon content less than 0.20%, manganese should be kept at the end of oxidation, and the manganese content should be adjusted to more than 0.20%. For steel grades containing Ni, Mo and W, the composition shall be adjusted to the lower specification limit or near the lower middle specification limit, and the content of other residual elements shall also meet the specification requirements.

2) 合适的化学成分。钢中的碳含量应达到所需要的范围：一般碳素钢应达到规格下限附近；碳素工具钢应达到规格的中下限，合金钢全扒渣的碳含量应加上铁合金带入的碳含量，再加上造渣材料和脱氧工艺的增碳量达到规格下限附近。磷含量在还原期只能增加，不能降低，这是由于全扒渣不彻底或飞扬悬挂在炉壁、炉盖处的渣中磷发生还原所致。另外，加入铁合金中的磷也被带入钢中。因此，氧化末期全扒渣前钢液中的磷含量应越低越好。一般扒渣前对各种规格的磷含量应符合表3-2中的规定。对于锰含量，在冶炼含锰低的钢种（如T10等钢）时，全扒渣前的锰含量应按表3-3的要求控制。对于用途重要的高级结构钢或碳含量（质量分数）低于0.20%的钢液，在氧化末期应保持锰，并将锰的含量（质量分数）调到0.20%以上。对于含Ni、Mo、W的钢种应将成分调到规格下限或中下限附近，其他残余元素的含量也应符合规格要求。

Table 3-2　Regulations of phosphorus content in steel before full slag removal at the end of oxidation

表3-2　氧化末期全扒渣前钢中磷含量（质量分数）的规定

Specification phosphorus content/% 规格磷含量/%	≤0.040	≤0.035	≤0.030	≤0.025	≤0.020
Phosphorus content before full slagging/% 全扒渣前磷量/%	≤0.020	≤0.015	≤0.012	≤0.010	≤0.008

Table 3-3　Manganese content control before full slagging at the end of oxidation

表3-3　氧化末期全扒渣前的锰含量（质量分数）控制

Specification manganese content/% 规格锰含量/%	0.15~0.30	0.15~0.35	0.20~0.40	0.35~0.60
Manganese content before full slagging/% 全扒渣前锰量/%	≤0.18	≤0.20	≤0.25	≤0.40

3) Adjust the fluidity of slag. The fluidity of slag should be adjusted before raking. Because the too thin slag will slip away from both sides of the rake head, it is also easy to bring out molten steel; the too thick slag is hard to operate and is not easy to pick out. Both of them can prolong the time of slagging and are not easy to clean. Therefore, it is very necessary to adjust the fluidity of slag before slagging.

3) 调整好熔渣的流动性。扒渣前要调整好熔渣的流动性。因过稀的熔渣会从耙头两侧溜开，也容易带出钢水；过稠的熔渣操作费力，也不易扒出。两者都延长扒渣时间，且又不易扒净。因此，扒渣前调整好熔渣的流动性也是十分必要的。

(2) Requirements and operation of full slagging at the end of oxidation. Because the content of FeO and phosphorus in the oxidation slag is very high, if it is not cleaned, it will be difficult to

deoxidize in the reduction period, the amount of deoxidizer will increase, the deoxidization time will be prolonged, and the phosphorus will be returned in the steel, so the full slag removal at the end of the oxidation period is required to be clean and thorough, and because the steel liquid is exposed in the process of slag removal, the temperature of the steel liquid will drop sharply and the suction will be serious, so the slag removal is required to be rapid. Therefore, before raking slag, it is necessary to pry off the slag at the furnace door and pad the furnace door (the slag removal of small and medium-sized furnace platforms usually passes through the furnace door), and prepare rake and slag tank in advance.

（2）氧化末期对全扒渣的要求及操作。因为氧化渣中（FeO）和磷含量很高，如果不扒净，还原期脱氧困难，脱氧剂用量增加，脱氧时间延长，同时钢中也回磷，所以氧化末期的全扒渣要求干净彻底，又因为扒渣过程中，钢水裸露，钢液急剧降温且吸气严重，所以又要求扒渣迅速。为此，扒渣前要撬掉炉门残渣并垫好炉门（中小型炉台的除渣一般都通过炉门），提前准备好耙子和渣罐等。

Rakes are usually made of wood or water-cooled. In order to avoid too much cooling of the liquid steel, most of them should be removed with electricity first, and then the electrode can be lifted slightly. The furnace body can also be poured to one side of the furnace door as required to facilitate the rapid removal of slag. In addition, on the large-scale furnace platform equipped with electromagnetic stirring, the stirring function can be used to gather the slag into the furnace door, so that the slag raking operation is easy to carry out.

扒渣多用木制或水冷的耙子。扒时应首先带电扒去大部分，然后方可略抬电极进行，以免钢液降温太多。也可根据需要向炉门一侧倾动炉体，以利于熔渣的快速扒除。此外，在装有电磁搅拌的大型炉台上，可利用搅拌作用，把熔渣聚集到炉门中处，使扒渣操作易于进行。

3.3.5.2 Carburization
3.3.5.2 增碳

Most of the carburization is caused by insufficient amount of decarburization or low final decarburization, which is an abnormal operation. It is easy to increase the content of gas and inclusions in the steel during carburization, which wastes raw materials and prolongs smelting time, so it should be avoided as much as possible.

增碳多是脱碳量不足或终脱碳过低所致，它是一种不正常的操作。因为增碳过程易使钢中气体和夹杂物含量增加，既浪费原材料，又延长冶炼时间，所以应尽量避免。

There are four common ways to increase carbon:

(1) Add pig iron and carbon. Because the method reduces the temperature of molten steel and requires the loading amount to be carried out under the condition that the molten pool allows, the amount of carburization is limited, generally not more than 0.05%.

(2) In general, it is not recommended to increase carbon consumption of electrode under power failure.

(3) Slagging for carbon. Although the yield of this method is not accurate enough, it is eco-

nomical and convenient, so it is more common.

(4) Dusting and carburizing. This method is rapid, simple, accurate and does not reduce the temperature of the molten pool, so it is the most ideal method for carbon addition at present.

常见的增碳方法有四种：

(1) 补加生铁增碳。由于该法降低钢液温度且又要求装入量在熔池允许的条件下进行，因此增碳量受到了限制，一般（质量分数）不大于 0.05%。

(2) 停电下电极增碳，但增加电极消耗，一般不提倡。

(3) 扒渣增碳。该法虽然收得率不够准确，但经济方便，因而比较多见。

(4) 喷粉增碳。该法操作迅速、简便，且准确而又不降低熔池温度，所以是目前最理想的增碳方法。

The amount of carbon brought in by iron content and the amount of carbon added by reduction process (such as carbide slag) shall be considered for carbon addition. In addition to pig iron, there are also electrode powder or coke powder in common use. Their yield is not only related to the quality and quantity of carburizer and carburizing method, but also related to the type of steel, smelting method, molten steel temperature and furnace life. Generally, the yield of electrode powder is higher than that of coke powder; the yield is lower when the amount of carburizing is more; the yield of medium carbon steel is higher than that of high carbon steel and low carbon steel. The higher the temperature of molten steel is, the higher the yield is; the higher the yield is, the higher the yield is.

增碳应考虑加入铁含量带入的碳量以及还原工艺的增碳量（如电石渣）。常用的增碳剂除生铁外，还有电极粉或焦炭粉等，它们的收得率不仅与增碳剂的质量和增碳数量及增碳方法有关，而且与所炼钢种、冶炼方法、钢液温度和炉龄情况有关。一般电极粉比焦炭粉收得率高；增碳量越多收得率越低；中碳钢比高碳钢和低碳钢的收得率要高；氧化法比不氧化法收得率要高；炉龄中期比炉龄初期和末期要低一些；钢液温度越高收得率越高；喷粉增碳的收得率比扒渣增碳高。

For slag raking and carburizing, in order to stabilize the yield of carburizing agent, the slag must be raked clean. After adding the carburizing agent, the rake shall be used to fully push and roll on the steel surface to promote the absorption of carbon powder by the liquid steel, and then the slag making material shall be added. In the process of slag raking and carburizing, in order to reduce suction and avoid inaccurate recovery of carburizing agent, it is better to turn off the power.

对于扒渣增碳，为了稳定增碳剂的收得率，必须将熔渣扒净，增碳剂加入后要用耙子在钢液面上进行充分地推搅，促进钢液对炭粉的吸收，然后加入造渣材料。扒渣增碳过程中，为了减少吸气和避免增碳剂的回收不准，最好不通电。

3.3.6 Oxidation Period Operation
3.3.6 氧化期操作

3.3.6.1 Main Indicators for Judging the Degree of Oxidation
3.3.6.1 判断氧化期进行程度的主要标志

In order to remove the gas and non-metallic inclusions in the steel better, the oxidation period must ensure that the molten pool has a certain intense boiling time, so a certain amount of decarburization and decarburization speed are required in the decarburization process. From the phenomenon point of view, the amount of decarburization, decarburization speed and intense boiling time are the main indicators to judge the degree of oxidation.

对于不配备炉外精炼的电炉炼钢,为了更好地去除钢中的气体和非金属夹杂物,氧化期必须保证熔池要有一定的激烈沸腾时间,因此在脱碳过程中要求要有一定的脱碳量和脱碳速度。从现象上看,脱碳量、脱碳速度、激烈沸腾时间就是判断氧化期进行程度的主要标志。

(1) Decarbonization amount. In the process of EAF steel production, the decarburization amount in the oxidation period is determined according to the requirements of the steel grades and technical conditions, smelting method and furnace charge quality. Generally speaking, the worse the charge quality is or the stricter the quality requirements for steel are, the higher the decarburization amount is required.

Production practice has proved that too little decarburization can not achieve the purpose of removing a certain amount of gas and inclusions in the steel; and too much decarburization does not significantly improve the quality of the steel, on the contrary, it will prolong the smelting time and increase the erosion of the furnace lining, and waste human and material resources, so too much decarburization is not necessary. It is generally believed that the decarburization amount of oxidation smelting is 0.20% ~ 0.40%, and the requirement of back blowing method is more than 0.10%, while the decarburization rate of small hearth is fast, so it can be specified slightly higher.

(1) 脱碳量。在电炉钢生产过程中,氧化期的脱碳量是根据所炼钢种和技术条件的要求、冶炼方法和炉料的质量等因素来确定。一般说来,炉料质量越差或对钢的质量要求越严,要求脱碳量越高些。

生产实践证明,脱碳量过少,达不到去除钢中一定量气体和夹杂物的目的;而脱碳量过大,对钢的质量并没有明显的改善,相反会延长冶炼时间及加重对炉衬的侵蚀,浪费人力物力,因此脱碳量过大也是没有必要的。一般认为,氧化法冶炼的脱碳量(质量分数)为0.20% ~ 0.40%,返吹法则要求大于0.10%,而小炉台因脱碳速度快,可规定略高一些。

(2) Decarbonization rate. Production practice has proved that decarburization speed is too slow, molten pool boiling is slow, which can not fully remove gas and inclusions; while decarburization speed is too fast, decarburization ends in a short time, which will inevitably cause molten pool boiling violently, which is easy to expose the liquid steel, suction is serious, and corrosion of furnace lining is aggravated, which is not only adverse to degassing and inclusion removal, but also causes accidents such as splashing and steel running. Therefore, the decarbonization of EAF steelmaking requires a certain speed. The proper decarburization speed should ensure that the degassing volume of molten steel in unit time is larger than the suction volume, and the inclusions can be fully discharged. Generally, the normal decarburization speed of ore is 0.008% ~ 0.015%/min, while that of oxygen blowing is 0.03% ~ 0.05%/min.

(2) 脱碳速度。生产实践证明,脱碳速度过慢,熔池沸腾缓慢,起不到充分去气除夹杂的作用;而脱碳速度过快,在短时间内结束脱碳,必然造成熔池猛烈的沸腾,易使钢液裸露，吸气严重，且对炉衬侵蚀加重，这不仅对去气除夹杂物不利，还会造成喷溅、跑钢等事故。所以，电炉炼钢的脱碳要求要有一定的速度。合适的脱碳速度应保证单位时间内钢液的去气量大于吸气量，并能使夹杂物充分排出。一般正常的矿石脱碳速度要求为 0.008%~0.015%/min，而吹氧脱碳速度要求为 0.03%~0.05%/min。

(3) Intense boiling time. The decarburization amount and decarburization speed in the oxidation period can't really reflect the boiling quality of the molten steel. The time of intense boiling of the molten pool must be considered again. Only in this way can the situation of degassing and inclusion removal in the steel and the uniform temperature of the molten steel be fully revealed. However, the time of intense boiling depends on the starting temperature of oxidation, slag condition and oxygen supply speed, that is, the time of intense boiling is directly related to the amount and speed of decarburization. However, even if the above factors are not completely available, the steelmaker of electric furnace can directly blow argon or CO gas into the molten pool to ensure that the molten pool has enough intense boiling time. It must be pointed out here that although argon or CO gas blown into the molten pool can produce good intense boiling, it can not solve the oxidation of impurities in steel, and the oxidation of many impurities is inseparable from oxygen. Therefore, if oxygen is not supplied to the molten pool during the oxidation period, the ideal pure molten steel can not be obtained by blowing argon or CO gas alone.

In the process of EAF steel production, the intense boiling time of the molten pool in the oxidation period should not be too short or too long. Generally, it can meet the requirements in about 15~20 minutes.

(3) 激烈沸腾时间。氧化期的脱碳量和脱碳速度往往还不能真实地反映钢液沸腾的好坏，必须再考虑熔池的激烈沸腾时间，只有这样才能全面地表明钢中去气除夹杂物及钢液温度的均匀情况等。然而熔池的激烈沸腾时间取决于氧化的开始温度、渣况及供氧速度等，即熔池的激烈沸腾时间与脱碳量和脱碳速度有直接关系。但即使上述因素不完全具备，电炉炼钢工也可通过直接向熔池吹入氩气或 CO 气体等，保证熔池具有足够的激烈沸腾时间。这里必须指出，向熔池中吹入氩气或 CO 气体，虽能制造良好的激烈沸腾，但它不能解决钢中杂质的氧化，而许多杂质的氧化是离不开氧的。因此，在氧化期如果不向熔池中供氧，单凭吹氩或吹一氧化碳气体，最终也不能获得较为理想的纯净钢液。

在电炉钢生产过程中，氧化期熔池的激烈沸腾时间不应过短或过长，一般在 15~20min 就可满足要求。

3.3.6.2 Oxidation Period Operation
3.3.6.2 氧化期操作

(1) Principles of operation in oxidation period. The tasks of oxidation stage are mainly completed by decarbonization. In terms of dephosphorization and decarbonization, both of them require the slag to have strong oxidation capacity, but dephosphorization requires medium low temperature, large slag volume and good fluidity, while decarbonization requires high temperature and thin slag, so the temperature of the molten pool is rising gradually. According to these characteristics, we summarize the general operation principles of the oxidation period as follows: in

terms of oxidation sequence, phosphorus comes first and then carbon; in terms of temperature control, slow comes first and then fast; in terms of slagging, large amount of slag goes first and then thin slag decarburizes; in terms of oxygen supply, ore or comprehensive oxidation can be carried out first, and finally oxygen blowing is the main way.

（1）氧化期操作的原则。氧化期的各项任务主要是通过脱碳来完成的。单就脱磷和脱碳来说，两者均要求熔渣具有较强的氧化能力，可是脱磷要求中等偏低的温度、大渣量且流动性良好，而脱碳要求高温、薄渣，所以熔池的温度是逐渐上升的。根据这些特点，将氧化期总的操作原则归纳如下：在氧化顺序上，先磷后碳；在温度控制上，先慢后快；在造渣上，先大渣量去磷，后薄渣脱碳；在供氧上，可先进行矿石或综合氧化，最后以吹氧为主。

（2）General operation during oxidation period. After the furnace charge is fully melted and stirred, samples shall be taken for analysis of C, Mn, S, P, Ni, Cr, Si and Cu. For example, Mo, W and other elements in steel shall also be analyzed. Then the slag is raked and new slag is made up, so that the slag content of the oxide slag is 3% ~ 4% of the material weight. In order to speed up the melting of slagging materials, oxygen can be used to blow the slagging layer. If the fluidity is not good, fluorite should be used for adjustment. When the temperature reaches above 1530℃, ore is used for oxidation. In the oxidation process, the decarburization speed should be controlled, and the time of intense boiling in the molten pool should be controlled; the decarburization amount should meet the process requirements, and if it is not enough, the appropriate time should be selected for carburization. In the oxidation process, it is best to be able to achieve automatic slag flow, which is not only conducive to dephosphorization, but also conducive to the later stage of thin slag carbon reduction. In order to know the decarburization and dephosphorization situation and the components of non oxidizing elements accurately, the relevant content should be analyzed in the oxidation process. When the ore is added or the blowing is stopped, the molten pool will enter into clean boiling, and some of them will keep manganese. At the same time, the composition of some non oxidized elements, such as Ni, Mo, etc., will be adjusted as required to reach the lower and middle limits of the specification, and then the composition of C, Mn, P and other relevant elements will be sampled and analyzed. When the molten pool has the conditions for slag removal, the full slag removal operation can be carried out, and then the reduction period can be transferred.

（2）氧化期的一般操作。炉料全熔经搅拌后，取样分析 C、Mn、S、P、Ni、Cr、Si、Cu，如钢中含有 Mo、W 等元素也要进行分析。然后扒渣并补造新渣，使氧化渣的渣量达到料重的 3%~4%。为了加速造渣材料的熔化可用氧气吹拂渣层，流动性不好时要用萤石调整，当温度达到 1530℃ 以上开始用矿石氧化。在氧化过程中，应控制脱碳速度，并掌握熔池的激烈沸腾时间；脱碳量要满足工艺要求，如果不足应选择适当时机进行增碳。在氧化过程中，最好能够做到自动流渣，这样既有利于脱磷，又有利于后期的薄渣降碳。为了掌握脱碳、脱磷情况及准确地知道不氧化元素的成分，在氧化中途还应分析有关的含量。当加完矿或停吹后，熔池进入清洁沸腾，有的还要保持锰，在这同时，根据需要还要调整一些不氧化元素的成分，如 Ni、Mo 等，使之达到规格的中下限，然后再取样分析 C、Mn、P 等及其他有关元素的成分。当熔池具备扒渣条件时，即可进行全扒渣操作，而后转入还原期。

In the oxidation process, the composition, fluidity and quantity of slag should be controlled correctly. Both dephosphorization and decarburization require high oxidation ability and good fluidity of slag. The ideal basicity of dephosphorization should be 2.5~3.0, while the basicity of decarbonization should be about 2.0. In the smelting process, the fluidity of oxide slag is deteriorated due to the collapse of furnace wall or the floating of large magnesia at furnace bottom, so it should be removed in time.

在氧化过程中,应正确控制熔渣的成分、流动性和渣量,无论是脱磷还是脱碳,都要求熔渣具有较高的氧化能力和良好的流动性。理想的脱磷碱度应保持为2.5~3.0,而脱碳的碱度为2.0左右。在冶炼过程中,有的因炉壁倒塌或炉底大块镁砂上浮,使氧化渣的流动性变差,这时应及时扒出。

Good oxidation slag should be foam slag, which can encircle arc light, which is conducive to the heating and protection of molten steel. After cooling, the surface is black and the fracture is dense and loose. This indicates that (FeO) content is high and alkalinity is suitable. At the end of oxidation, the slag sometimes thickens, which is mainly caused by the powder magnesia and a large number of non-metallic inclusions floating up. When smelting high carbon steel, if the slag is dry and the surface is rough and light brown, it indicates that (FeO) content is low and the oxidation performance is poor. This phenomenon appears more in the back blowing smelting or pure oxygen oxidation. When smelting low carbon steel, if the oxide slag surface is black and bright, and the slag is very thin, it indicates that (FeO) content is high and alkalinity is low, lime should be added at this time.

良好的氧化渣应是泡沫渣,可包围住弧光,从而有利于钢液的升温和保护炉衬,冷却后表面呈油黑色,断口致密而松脱,这表明(FeO)含量较高、碱度合适。氧化末期有时氧化渣发稠,这主要是炉衬粉化的镁砂和大量的非金属夹杂物上浮造成的。冶炼高碳钢时,如熔渣发干,表面粗糙且呈浅棕色,表明(FeO)含量低,氧化性能差,这种现象在返吹法冶炼或纯氧氧化时出现较多。冶炼低碳钢时,如氧化渣表面呈黑亮色,渣又很薄,表明(FeO)含量高,碱度低,这时应补加石灰。

(3) Several typical operations in oxidation period. The oxidation period starts under the condition that the composition and temperature of the melt cleaning are confirmed to be appropriate, but sometimes the composition of the melt cleaning is not so ideal. The common typical situations are high carbon and high phosphorus; high carbon and low phosphorus; low carbon and high phosphorus; low carbon and low phosphorus, etc. as follows:

(3) 氧化期常见的几种典型操作。氧化期是在确知熔清成分和温度合适的条件下开始的,但有时熔清成分不是那么理想,常见的几种典型情况有:碳高、磷高;碳高、磷低;碳低、磷高;碳低、磷低等,现分述如下:

1) High carbon and phosphorus. At this time, in the early stage of oxidation, the opportunity of low temperature of molten pool should be used to concentrate on dephosphorization, and in the process of dephosphorization, the temperature should be gradually increased to create conditions for later decarbonization. The specific operation is: after fully melting and raking the slag, a large amount of slag can be produced. The oxidized slag can be blown and heated up, and

then ore powder or iron oxide scale and a certain amount of ore can be added to facilitate dephosphorization. At the same time, to ensure the good fluidity of the slag, when the temperature is appropriate, add the ore in batches to make decarburization boiling, and automatically flow the slag, supplement the new slag or carry out the slag change operation, so that the phosphorus can meet the permissive conditions for slag raking very soon. It is better to use powder spray for dephosphorization after full melting. In the case of high carbon and phosphorus, dephosphorization is the main operation in the early stage of oxidation, and decarbonization is the main operation in the later stage. Of course, the high-level operation can also take both into account until the process requirements are fully met.

1) 碳高、磷高。此时应在氧化初期,利用熔池温度偏低的机会集中力量脱磷,并在脱磷过程中,逐渐升温,为后期脱碳创造条件。具体操作是:全熔扒渣后制造较大的渣量,可吹氧化渣并升温,然后加入矿石粉或氧化铁皮及适量的矿石,以利于脱磷。在这同时,要保证熔渣流动性良好,当温度合适后,再分批加入矿石制造脱碳沸腾,并自动流渣,补充新渣或进行换渣操作,这样很快就能使磷满足扒渣的许可条件。如果全熔换渣后改用喷粉脱磷效果更好。在碳高、磷高的情况下,氧化前期的操作以脱磷为主,后期以脱碳为主。当然,高水平的操作也可两者兼顾,直至全面满足工艺要求为止。

2) High carbon and low phosphorus. If the permissive conditions for slagging are not met, the operation at this time can not only use powder spraying for dephosphorization, but also use decarburization boiling and dephosphorization in the process of decarburization. The specific operation is: after full melting and slagging, a large amount of slags will be produced, the oxygen slags will be used for heating up, when the temperature is appropriate, the carbon will be reduced, and the ore or iron oxide scale will be added in time, so that the slags will flow automatically, and the new slags will be supplemented, so that the phosphorus can be reduced to the conditions permitted by slagging, and the high temperature and thin slags will be used for decarbonization until the process requirements are met.

2) 碳高、磷低。如果没有满足扒渣的许可条件,这时的操作除采用喷粉脱磷外,还可利用脱碳沸腾,并在脱碳过程中去磷。具体操作是:全熔扒渣后制造较大的渣量,用氧气化渣升温,当温度合适后开始降碳,并适时地加入矿石或氧化铁皮等,使其自动流渣、并补选新渣,这样很快就能将磷降到扒渣许可的条件,最后采用高温、薄渣脱碳直至满足工艺要求为止。

If the permissive conditions for slagging have been met, the operation at this time is mainly to make the molten pool boiling and reduce carbon, while heating up. The specific operation is as follows: after the slag is fully melted and raked, the appropriate slag volume is manufactured, and the slag is rapidly melted and heated up with oxygen. When the temperature is appropriate, mineral oxygen can be used in combination or pure oxygen can be used for decarburization until the process requirements are met.

如果已满足扒渣的许可条件,这时的操作主要是制造熔池沸腾和降碳,与此同时升温。具体操作是:全熔扒渣后制造合适的渣量,用氧气快速化渣与升温,当温度合适后,可采用矿氧并用或纯氧脱碳,直至满足工艺要求为止。

3) Low carbon and high phosphorus. At this time, it is necessary to concentrate on dephosphorization, and then increase carbon. When the temperature is up, the decarburization boiling can be manufactured until the process requirements are met.

3) 碳低、磷高。这时应集中力量去磷，然后增碳，当温度上来后，再制造脱碳沸腾直至满足工艺要求为止。

4) Low carbon and phosphorus. At this time, the operation is mainly to increase carbon, and then decarburization is intense boiling, while rapid heating. If the charge quality is good, i.e. the impurities are less, and the time of intense boiling is not enough, the boiling of the molten pool can also be compensated by blowing argon or CO gas directly to meet the process requirements.

4) 碳低、磷低。这时的操作主要是增碳，然后再脱碳激烈沸腾，与此同时快速升温。如果炉料质量较好，即杂质较少，而激烈沸腾时间不够时，也可借助于直接吹入氩气或吹一氧化碳气体来弥补熔池的沸腾，以满足工艺要求。

Exercises
（1）What is the main task of oxidation period?
（2）What are the oxidation methods?
（3）What are the conditions for full slagging at the end of oxidation?
（4）What is the operation principle of oxidation period?

思考题
（1）氧化期的主要任务是什么？
（2）氧化方法有哪些？
（3）氧化末期的全扒渣条件是什么？
（4）氧化期的操作原则是什么？

Task 3.4　Restore Period Operation
任务3.4　还原期操作

Mission objectives
任务目标
（1）Understand deoxidation operation in reduction period.
（1）了解还原期脱氧操作。
（2）Master the deoxidation method of electric furnace steelmaking.
（2）掌握电炉炼钢脱氧方法。
（3）Understand the desulfurization operation in reduction period.
（3）了解还原期脱硫操作。

The specific tasks of reduction refining are as follows:

(1) Remove oxygen from molten steel as much as possible.

(2) Remove sulfur from molten steel.

(3) Finally adjust the chemical composition of liquid steel to meet the specification requirements.

(4) Adjust the temperature of molten steel and create conditions for normal pouring of steel.

还原精炼的具体任务是:

(1) 尽可能脱除钢液中的氧。

(2) 脱除钢液中的硫。

(3) 最终调整钢液的化学成分,使之满足规格要求。

(4) 调整钢液温度,并为钢的正常浇注创造条件。

The above tasks are interrelated and carried out at the same time. The deoxidation of molten steel is good, which is good for desulfuration, stable chemical composition and high yield of alloy elements, so deoxidation is the key link of reduction refining operation.

上述任务的完成是相互联系,同时进行的。钢液脱氧好,有利于脱硫,且化学成分稳定,合金元素的收得率也高,因此脱氧是还原精炼操作的关键环节。

3.4.1 Deoxidation
3.4.1 脱氧

3.4.1.1 Formation and Elimination of Deoxidation Products
3.4.1.1 脱氧产物的形成与排除

(1) Formation of deoxidation products. The commonly used deoxidizers for EAF steelmaking are C, Mn, Si, Al and calcium alloy. Except for the escape of CO gas generated by the reaction of carbon and oxygen, the deoxidizing products of other elements in the molten steel mainly exist in the form of silicate or aluminate. Therefore, the deoxidizing products described here mainly refer to the oxide system inclusions in the steel.

(1) 脱氧产物的形成。电炉炼钢常用的脱氧剂有 C、Mn、Si、Al 及钙系合金等,其中除碳与氧反应生成 CO 气体逸出外,其他各种元素在钢液中的脱氧产物主要是以硅酸盐或铝酸盐形式存在,因此这里所叙述的脱氧产物主要是指钢中的氧化系夹杂。

The formation of deoxidation products is composed of nucleation and growth, which is also the first step of deoxidation process. For Al, Zr and Ti elements with strong deoxidizing ability, the probability of homogeneous nucleation is high due to the large supersaturation and energy fluctuation in the micro volume. For Si and Mn elements with weak deoxidizing ability, it is possible to nucleate on the ready-made matrix in the melt, such matrix always exists in the molten steel, such as inclusions or other atomic groups or concentration differences and other different interfaces, so the nucleation of deoxidizing products in the molten pool is generally easier.

脱氧产物形成由成核和长大两个环节组成,这也是脱氧过程的首要步骤。对于脱氧能力很强的 Al、Zr、Ti 元素,由于在微观体积内具有较大的过饱和度和能量起伏,所以均相成核的概率性较大。而对于脱氧能力较弱的 Si、Mn 元素,有可能依附在熔体内的现成基体上成核，这样的基体在钢液中总是存在的，如夹杂物、其他原子集团、浓度差及其他不同的界面等，因此脱氧产物在熔池中的成核一般是比较容易的。

Once nucleation occurs, the concentration of the surrounding deoxidizer and oxygen will immediately decrease. In order to maintain the balance of the concentration, these elements will continue to diffuse from a distance, resulting in nuclear growth. It has been calculated that it takes only 0.2s to grow up to 90% of the final radius, and only 12.8s to grow up to 90% of the final radius for the deoxidation product with the final radius of 20μm. It can be seen that the growth of the deoxidized products is also very fast. The deoxidizing products will inevitably float up under the action of interface tension or agitation, which may be eliminated.

成核一旦发生，周围的脱氧剂和氧的浓度就立刻降低，为保持浓度的平衡，这些元素将不断地从远处扩散过来，从而引起核的长大。有人曾计算溶解氧从 0.06% 降到 0.01% 的硅脱氧，最终半径为 2.5μm 的脱氧产物，长大到最终半径的 90% 只需 0.2s，最终半径为 20μm 的脱氧产物长大到 90% 也只需 12.8s，可见脱氧产物的长大也是很快的。形成的脱氧产物因比钢轻及在界面张力或搅拌等因素的作用下必然引起上浮，从而有可能排除。

(2) Elimination of deoxidation products and its influencing factors. The removal degree of deoxidized products from steel mainly depends on their floating speed in molten steel, and the floating speed is related to the composition, shape, size, melting point, density of deoxidized products, interfacial tension, viscosity of molten steel and agitation, etc., and roughly obeys Stokes formula:

$$v = K\frac{2r^2 g}{9\eta}(\rho' - \rho) \tag{3-1}$$

Where　　v——floating Velocity of deoxidizing product particles, cm/s;

r——particle inclusion radius, cm;

g——acceleration of gravity, cm/S^2;

ρ'——density of liquid steel, g/cm^3;

ρ——density of deoxidation products, g/cm^3;

η——viscosity of liquid steel, Pa·s;

K——constant, $K=1$ can be selected when deoxidizing product floats up in molten steel.

(2) 脱氧产物的排除及其影响因素。脱氧产物从钢中的去除程度主要取决于它们在钢液中的上浮速度，而上浮速度又与脱氧产物的组成、形状、大小、熔点、密度以及界面张力、钢液的黏度与搅拌等因素有关，并大致服从斯托克斯公式：

$$v = K\frac{2r^2 g}{9\eta}(\rho' - \rho) \tag{3-1}$$

式中　　v——脱氧产物颗粒夹杂上浮速度，cm/s;

r——颗粒夹杂半径，cm;

g——重力加速度，cm/s^2;

ρ'——钢液密度，g/cm^3;

ρ——脱氧产物密度，g/cm^3;

η——钢液黏度，Pa·s;

K——常数，脱氧产物在钢液中上浮时可选用 $K=1$。

It can be seen from the formula (3-1) that reducing the viscosity of liquid steel is beneficial to the floating of particle inclusions. However, the viscosity of liquid steel is related to composition and temperature. It is limited to reduce the viscosity of liquid steel by adjusting composition, while the viscosity of liquid steel changes little with temperature. For example, when the temperature of 30 steel rises from 1535℃ to 1610℃, the viscosity only decreases by 0.00055Pa·s. It can be seen from the calculation that for the steel with certain chemical composition, the floating speed can only be increased by 3 times under the most ideal conditions by increasing the temperature to improve the fluidity of the liquid steel. It can also be seen from the formula (3-1) that the larger the density difference is, the more favorable it is for the upward floating of particle inclusions. However, the density range of different oxides in the molten steel is small, and the upward floating speed can be increased by 2～3 times at most by changing the particle density. Therefore, the removal of oxide inclusions in steel mainly depends on increasing its radius. In this case, Stokes formula can be simplified as follows:

$$v = Kr^2 \quad (3-2)$$

由式 (3-1) 可看出，降低钢液黏度有利于颗粒夹杂的上浮。但钢液的黏度与成分、温度有关，依靠调整成分来降低钢液的黏度是有限的，而钢液的黏度随着温度的变化不大，例如 30 钢当温度由 1535℃ 升至 1610℃ 时，黏度仅降低 0.00055Pa·s。通过计算可知，对于化学成分一定的钢种，提高温度改善钢液的流动性，在最理想的条件下，上浮速度只能增加 3 倍左右。从式 (3-1) 还可看出，密度差越大，越有利于颗粒夹杂的上浮，但钢液内不同氧化物的密度变化范围较小，依靠改变颗粒密度至多能将上浮速度提高 2～3 倍。因此，要去除钢中氧化系夹杂，主要是依靠增加它的半径。此时，斯托克斯公式可简写成式 (3-2)：

$$v = Kr^2 \quad (3-2)$$

The formula (3-2) shows that v is directly proportional to the square of r. Therefore, the larger the particle radius of deoxidized product is, the faster the floating speed is. For example, in the same condition of molten steel, when the inclusion particle radius r increases from 20μm to 40μm, the floating speed increases by four times. Because the physical and chemical properties of deoxidized products, interfacial tension and inhomogeneity of molten steel temperature are not considered in the above discussion, and various dynamic conditions of melt are not considered, the calculation results are about 2～4 orders of magnitude smaller than the actual observation. Therefore, Stokes formula can only qualitatively estimate the influence of the increase of deoxidation product radius on the floating velocity, but it has limitations in calculation.

式 (3-2) 表明，v 与 r 的平方成正比。因此，脱氧产物的颗粒半径越大，上浮速度越快，例如在同一条件下的钢液中，当夹杂物颗粒半径 r 由 20μm 增加到 40μm 时，上浮速度提高 4 倍。上述讨论由于没有考虑脱氧产物的物化性质、界面张力以及钢液温度的不均匀性，也没有考虑熔体所处的各种动力学条件等，因此计算结果与实际观察小 2～4 个数量级。所以，斯托克斯公式只能定性地估计脱氧产物半径的增大对上浮速度的影响，而用于计算有局限性。

The aggregation and growth process of oxide inclusions suspended in molten steel is called agglomeration process. The coalescence process is spontaneous and is accomplished by reducing the polymerization force produced by surface free energy. However, it is easier for liquid and viscous deoxidation products to coalesce and float up than solid and viscous particles, which is determined by their different physical characteristics. Therefore, the selection of deoxidizing method and deoxidizer is considered as follows: the concentration of dissolved oxygen in steel is minimized, and deoxidizing products with low melting point and good fluidity are produced. However, although the solid and viscous deoxidizing products are difficult to coalesce, under certain conditions, as long as there is a chance of contact and collision and the surface free energy is reduced, they will coalesce more and more, and accelerate to float up.

悬浮于钢液内氧化物夹杂的聚集、长大过程称为聚结过程。聚结过程是自发的,并通过降低表面自由能所产生的聚合力来完成。然而液态的、黏性小的脱氧产物比固态的、黏性大的颗粒聚结上浮更容易,这是由它们各自不同的物理特性决定的。因此,脱氧方法和脱氧剂的选择是这样考虑的:最大限度地降低钢中溶解氧的浓度,并生成低熔点、流动性好的脱氧产物。然而,尽管固态的、黏性大的脱氧产物聚结比较困难,但在一定的条件下,只要有机会接触碰撞且通过表面自由能的降低,也会越聚越大,并加速上浮。

Since the liquid deoxidizing products in molten steel are easy to coalesce and float, what kind of deoxidizing products are liquid? The melting point of the individual deoxidizing products of various elements is higher than the steel-making temperature, and the stronger the deoxidizing ability of elements is, the higher the melting point of deoxidizing products, such as Al_2O_3, TiO_2, SiO_2, etc. Such deoxidizing products exist as solid particles at steelmaking temperature. In the process of deoxidization, the liquid deoxidizing products with low melting point, such as $MnO \cdot SiO_2$, $2MnO \cdot SiO_2$, $MnO \cdot Al_2O_3$ and so on, are easy to be formed. It is not difficult to see that the melting point of deoxidizing products also has a great influence on its exclusion from steel.

既然钢液中呈液态的脱氧产物易于聚结上浮,那么什么样的脱氧产物呈液态呢?各种元素的单独脱氧产物的熔点都高于炼钢温度,且元素的脱氧能力越强,脱氧产物的熔点越高,如 Al_2O_3、TiO_2、SiO_2 等。类似这样的脱氧产物在炼钢温度下都以固体颗粒状态存在。而复合脱氧剂或同时加入几种单元素的脱氧剂,在脱氧过程中易于生成低熔点的液态脱氧产物,如 $MnO \cdot SiO_2$、$2MnO \cdot SiO_2$、$MnO \cdot Al_2O_3$ 等。由此不难看出,脱氧产物的熔点对其从钢中的排除也有很大的影响。

The interfacial tension between deoxidized products and molten steel also has a great influence on the floatation and removal of deoxidized products. In addition, due to the different composition of deoxidized products, the interfacial tension between them and molten steel is different, so it also affects the floatation and removal. For example, the inclusions of low carbon steel contain about 95% liquid inclusions of FeO and MnO. The measured interfacial tension between them and the molten steel is 0.175~0.3N/m. When Al_2O_3 is added to the melt, the interfacial tension can surge to 1.2N/m. Al_2O_3 in inclusions. The less the content and the lower the interfacial tension, the more difficult they are to be excluded. It has been proved that the stronger the chemical bond, the higher the melting point and the greater the polymerization force of the deoxidized products, the weaker their chemical interaction with the molten steel. Therefore, even if the refractory solid particles have enough interfacial tension, they can float up from the steel to the slag layer at a very fast speed, that is to say, the deoxidizing products with many strong deoxidizers are easy to float up and get rid of. For example, when the amount of aluminum used for bearing steel is increased from 0.22kg/t to 2kg/t, the amount of oxide inclusions in the steel is decreased from 0.011% to 0.004% in an instant. The main reason why Al_2O_3 can float up and get rid of quickly is the great polymerization force.

脱氧产物与钢液间的界面张力大小对脱氧产物的上浮与排除也有很大的影响。另外，因脱氧产物的组成不同，它们与钢液间的界面张力不同，所以也影响上浮与排除。例如，低碳钢的夹杂物中，大约含有（质量分数）95%的 FeO 和 MnO 的液态夹杂，它们与钢液间的界面张力实测为 0.175~0.3N/m，当向此熔体中加入 Al_2O_3 时，界面张力约能激增到 1.2N/m。夹杂物中 Al_2O_3 的含量越少，界面张力越低，它们也就越不易排除。研究证明，脱氧产物的化学键越强，熔点越高，聚合力也越大，它们同钢液的化学作用也就越弱。因此，即使难熔的固体颗粒只要具有足够的界面张力，也能够以很快的速度从钢中上浮到渣层，也就是说，有许多强脱氧剂的脱氧产物也容易上浮与排除。如冶炼轴承钢，当用铝量由 0.22kg/t 钢增至 2kg/t 钢，钢中氧化物夹杂量由 0.011% 瞬间降至 0.004%，Al_2O_3 能够很快地上浮与排除的原因主要是聚合力很大。

Under the condition of certain chemical composition, increasing temperature and decreasing viscosity of molten steel are beneficial to the floatation of deoxidized products. Temperature has a great influence on the floatation and removal of deoxidized products. When other conditions are the same, pure steel can be obtained by high temperature smelting. This is because high temperature can not only improve the fluidity of molten steel, but also liquefy some solid granular deoxidizing products, which is conducive to coalescence, floatation and removal. However, the smelting temperature should not be too high, because too high temperature not only increases the power consumption, but also affects the smooth operation of pouring. In addition, it is easy to reintroduce unnecessary foreign inclusions, which worsens the quality of steel.

在化学成分一定的情况下，提高温度降低钢液黏度有利于脱氧产物的上浮。温度对脱氧产物上浮与排除的影响是很大的。当其他条件相同时，高温冶炼能够获得较纯净的钢。这是因为高温除能改善钢液的流动性外，还能使一些固态的颗粒状脱氧产物得到相应的液化，有利于聚结、上浮与排除。然而冶炼的温度又不能过高，因为过高的温度，不仅增加电耗，也影响浇注工作的顺利进行。此外，高温吸气及冲刷、侵蚀耐火材料严重，容易重新引进不必要的外来夹杂物，进而又恶化了钢的质量。

Stirring can make the molten steel produce turbulence, increase the collision probability of deoxidizing products and accelerate the coalescence and floating speed, which is beneficial to the elimination of deoxidizing products. In addition to manual stirring, mechanical stirring, electromagnetic induction stirring and gas stirring in front of the furnace, powder spraying and argon blowing smelting are also popular in the ladle at present, especially after tapping. While deoxidizing and desulfurizing, the activity of deoxidizing product SiO_2 can be reduced, which is more conducive to their removal.

搅拌可使钢液产生紊流运动,使脱氧产物的碰撞几率增多及聚结且上浮速度加快,从而有利于脱氧产物的排除。炉前除了采用人工搅拌、机械搅拌、电磁感应搅拌、气体搅拌外,目前在包中进行喷粉与吹氩冶炼也盛行起来,尤其是出钢后的喷粉操作,在进行脱氧、脱硫的同时,还可降低脱氧产物 SiO_2 等的活度,更有助于它们的排除。

In short, the degree of elimination of deoxidized products in steel depends on the composition and properties of deoxidized products in the smelting process, which is directly related to the deoxidation process. The larger the particle size of the deoxidized product, or the greater the density difference, or the lower the melting point, the liquid and the greater the interfacial tension between the deoxidized product and the molten steel, the better the degree of elimination. In addition, the control of smelting temperature and the enhancement of stirring in different forms are also conducive to the floatation and removal of deoxidized products.

总之,钢中脱氧产物的排除程度,在冶炼过程中取决于脱氧产物的组成和性质,这与脱氧工艺有直接关系。脱氧产物的颗粒越大,或密度差越大,或熔点低呈液态并与钢液间的界面张力越大,排除程度越好。此外,控制合适的冶炼温度及加强不同形式的搅拌,也有利于脱氧产物的上浮与排除。

3.4.1.2 Deoxidation Method of EAF Steelmaking
3.4.1.2 电炉炼钢的脱氧方法

(1) Direct deoxidation. Direct deoxidation is the direct action of deoxidizer and molten steel. It can be divided into precipitation deoxidization and powder spraying deoxidization. After cleaning the oxide slag, the massive deoxidizer, such as ferromanganese, silicon manganese alloy or aluminum block (cake) or other multi-element deoxidizer, is put into (inserted) steel directly or added to the mirror surface of the molten steel, and then the thin slag is reduced. This deoxidization method is called precipitation deoxidization of molten steel. This concept originated from the precipitation reaction, and the process of oxide precipitation from molten steel belongs to the precipitation reaction. The precipitation deoxidization speed of molten steel is faster and the reduction time can be shortened, but the deoxidization products are easy to remain in the steel and become inclusions. In the process of deoxidization of molten steel, a special deoxidizing powder is delivered to the molten steel by means of metallurgical spraying device and inert gas (argon) as carrier. Because the specific surface area of deoxidizing powder (the ratio of the boundary area between deoxidizing powder and liquid steel to the volume of liquid steel) is several orders of magnitude larger than that of static slag steel interface under the condition of blowing, and under the strong agitation of current carrying argon, the diffusion mass transfer coefficient and the kinetic conditions of reaction are increased, so the deoxidizing speed of liquid steel is very fast, that is, in a very short time the deoxidation task can be completed well, the smelting process can be simplified, the refining time can be shortened, and various consumption can be reduced. In addition, the deoxidizer with low density, low boiling point or high vapor pressure (such as Ca, Mg, etc.) has been widely used in the deoxidization of molten steel. At the same time, the prop-

erty, shape, quantity and distribution of inclusions in the steel can be changed, so that the mechanical properties and technological properties of the steel can be improved.

（1）直接脱氧。直接脱氧就是脱氧剂与钢液直接作用，它又分为沉淀脱氧和喷粉脱氧两种。扒净氧化渣后，迅速将块状脱氧剂，如锰铁、硅锰合金或铝块（饼）或其他多元素的脱氧剂，直接投入（插入）钢中或加到钢液的镜面上，然后造还原稀薄渣，这种脱氧方法称为钢液的沉淀脱氧。这个概念起源于沉淀反应，因凡从钢液中析出氧化物的过程多属于沉淀反应过程。钢液的沉淀脱氧的速度较快，可缩短还原时间，但脱氧产物易残留在钢中而成为夹杂。钢液的喷粉脱氧是将特制的脱氧粉剂，利用冶金喷射装置并以惰性气体（氩气）为载体输送到钢液中去。由于在喷吹的条件下，脱氧粉剂的比表面积（脱氧粉剂和钢液间的界面积与钢液的体积比）比静态渣钢界面的比表面积大几个数量级，以及在载流氩气的强烈搅拌作用下，增大了扩散传质系数和改善了反应的动力学条件，因此钢液的脱氧速度很快，即在极短的时间内就可较好地完成脱氧任务，进而简化了冶炼工艺，缩短精炼时间，且又能降低各种消耗。另外，钢液的喷粉脱氧使密度小、沸点低或在炼钢温度下蒸气压很高的强脱氧剂（如Ca、Mg等）获得了广泛的应用；同时又可改变钢中夹杂物的属性和形态、数量与分布等，从而使钢的力学性能及工艺性能得到了提高。

At present, there are two ways of deoxidization of molten steel: one is in the furnace, the other is in the ladle. Because of the shallow molten pool and serious splashing, the deoxidizing powder is easy to escape with the current carrying bubble and burn on the slag surface, so the utilization rate of deoxidizing powder is low, but the final deoxidizing effect is better than the precipitation deoxidizing of molten steel, rather than the powder deoxidizing in ladle. The powder spray deoxidization in the bag has a high utilization rate due to the long travel of powder movement. Under the action of argon agitation, the probability of collision and coalescence of deoxidized products is high, and it is easy to float up and remove, while a few inclusions remain in the steel, which are also small, dispersed, evenly distributed, or their properties and morphology have changed, so the damage to the steel is relatively small. In addition, there is no secondary oxidation in the deoxidization of powder injection in ladle. There are many deoxidizers for spraying powder into molten steel. In addition to calcium alloy powder, there are also aluminum powder, ferrosilicon powder, ferrotitanium powder, rare earth, etc. slag powder can also be injected, such as CaO powder or a small amount of CaF_2 powder, or mixture of slag powder and some deoxidizing elements. These powders shall be strictly baked and screened before spraying. No matter in furnace or in ladle, the deoxidization effect is different, but the reaction principle is unchanged.

目前，钢液的喷粉脱氧方式有两种：一种是在炉内进行；另一种是在钢包中进行。炉内的喷粉脱氧因熔池浅，喷溅严重，脱氧粉剂容易随着载流气泡逸出并在渣面上燃烧，所以脱氧粉剂的利用率偏低，但最终脱氧效果还是强于钢液的沉淀脱氧，而不如钢包中的喷粉脱氧。包中的喷粉脱氧，由于粉剂运动的行程长，因此利用率很高。脱氧产物在氩气搅拌作用下，碰撞聚结几率大，易于上浮与排除，而少数夹杂物就是残留在钢中，也是细小、分散、均匀分布，或属性与形态发生了改变，因此对钢的危害也较小。此外，钢包中的喷粉脱氧无二次氧化。钢液的喷粉脱氧剂有多种，除钙系合金粉剂可用来脱氧外，还有铝粉、硅铁粉、钛铁粉、稀土等,也可喷吹渣粉,如 CaO 粉或掺入少量的 CaF$_2$ 粉,也可喷吹渣粉和某些脱氧元素的混合剂。这些粉剂喷前需经严格烘烤、筛分。钢液的喷粉脱氧无论是在炉内进行,还是在钢包中进行,只是脱氧效果有差异,而反应机理不变。

The deoxidization of slag powder (such as 85% CaO and 15% CaF$_2$) is not the direct effect of slag powder and oxygen in steel, but after the slag powder is injected, the a_{SiO_2} is reduced, the deoxidization ability of silicon is enhanced, and the oxygen content in steel is reduced. It is also believed that under the condition of blowing, the slag powder melts on the surface of the argon bubble to form a layer of liquid slag film, which makes silicon and oxygen in the steel diffuse to the slag film together, thus generating deoxidizing products with low activity and discharging with the argon bubble, thus further reducing the oxygen in the steel. Therefore, enough silicon must be contained in the molten steel when slag powder is used for deoxidization. In addition, the temperature drop of this method is large, and the hydrogen and silicon are increased, which can not be ignored in the formulation of process system. Because the specific surface area of deoxidizer is large, some of deoxidizer contact with molten steel in gas state, and diffusion and mass transfer are good under the stirring effect of argon, so the deoxidization of molten steel by powder injection has its own unique dynamic conditions. In particular, the oxygen in steel can be reduced to below $20 \times 10^{-4}\%$ by accurate operation of powder injection deoxidization in ladle.

喷吹渣粉[如喷吹 $w(CaO) = 85\%$ 和 $w(CaF_2) = 15\%$]的脱氧,并不是渣粉与钢中氧直接作用,而是渣粉喷入后,降低了 a_{SiO_2},增强了硅的脱氧能力,从而降低了钢中的氧含量。也有人认为在喷吹条件下,渣粉在氩气泡表面熔化形成一层液体渣膜,使钢中硅和氧一起向渣膜扩散,从而生成了活度低的脱氧产物并随同氩气泡排出,而使钢液中的氧得到了进一步的降低。因此,利用喷吹渣粉脱氧,钢液中必须含有足够的硅量。此外,该法温降大且增氢降硅,这在制定工艺制度时不可不考虑。在喷吹时,由于脱氧粉剂的比表面积大,有的还以气态与钢液接触,以及在氩气的搅拌作用下,扩散传质好,因此,钢液的喷粉脱氧有其独特的动力学条件。尤其是钢包中喷粉脱氧,准确的操作可将钢中氧降到 $20 \times 10^{-4}\%$ 以下。

The main factors influencing the deoxidization of molten steel by powder injection are as follows:

1) Injection parameters. The position, depth and angle of the injection gun have a strong influence on the size of the reaction interface and the residence time of the powder in the molten steel. The appropriate powder spraying intensity and mixing concentration can increase the injection amount and increase the intensity of slag steel mixing. However, it is unnecessary to increase the powder feeding speed. Because the excessive feeding speed makes it difficult for the deoxidizer to fully react with the molten steel and to escape early or expose the molten steel seriously, the best spraying process parameters should be selected according to the inherent characteristics of the powder spraying equipment.

影响钢液喷粉脱氧的主要因素有:

1) 喷吹参数。喷枪插入的位置、深度及角度等有力地影响反应界面的大小和粉剂在钢液内的停留时间,直接影响脱氧效果。合适的喷粉强度和混合浓度能相应增加喷入量,提高渣钢的激烈搅拌程度。但过大的送粉速度也是不必要的，因为过大的送粉速度使脱氧粉剂难以和钢液充分作用而过早的逸出或使钢液裸露严重，因此应根据喷粉设备的固有特点选择最佳的喷粉工艺参数。

2) Injection amount and injection time of deoxidizing powder. Generally, the larger the injection amount of deoxidizing powder is, the lower the final oxygen content in the molten steel; under the condition of certain injection intensity, the longer the injection time is, the higher the deoxidizing efficiency is. In addition, the cleaning time of argon blowing for molten steel after spraying should not be too short, otherwise, the deoxidized products will not be removed. However, if the time of argon blowing and washing is too long, it is easy to cause serious erosion of lining and spray gun, and the molten steel will be oxidized again, so the long time of argon blowing and washing is unnecessary.

2) 脱氧粉剂喷入量和喷吹时间。一般是脱氧粉剂的喷入量越大，钢液中的最终氧含量越低；在喷吹强度一定的条件下，喷吹时间越长，脱氧效率越高。此外，喷后对钢液的吹氩洗涤时间不应过短，否则脱氧产物来不及排除，但喷吹和洗涤时间过长，易使包衬和喷枪侵蚀严重，且钢液又会重新氧化，因此过长的喷吹和洗涤时间也是不必要的。

3) Deoxidation products. The products of deoxidization by spraying molten steel powder have a great influence on deoxidization effect. If the deoxidized product is a large spherical fusible inclusion (such as $CaO-Al_2O_3$), it can float up and be eliminated quickly.

3) 脱氧产物。钢液喷粉脱氧的产物对脱氧效果的影响很大。如果脱氧产物为大型球状易熔夹杂（如 $CaO-Al_2O_3$），就能很快地上浮与排除。

4) Lining material. A large number of scientific experiments and production practices have proved that SiO_2 in clay brick lining will react with Ca and Al in the process of blowing:

$$SiO_{2(M)} + 2\{Ca\} = 2(CaO) + [Si]$$
$$3SiO_{2(M)} + 4[Al] = 2(Al_2O_3) + 3[Si]$$

4) 包衬材质。大量的科学实验和生产实践已经证明，在喷吹过程中，黏土砖包衬中的 SiO_2 将与 Ca 和 Al 发生下述反应：

$$SiO_{2(M)} + 2\{Ca\} = 2(CaO) + [Si]$$
$$3SiO_{2(M)} + 4[Al] = 2(Al_2O_3) + 3[Si]$$

It can be seen that the erosion of clay brick lining is serious in the process of blowing, and a large amount of SiO_2 is reduced, which increases the Si content in steel, affects the utilization ratio of Ca and Al, and makes the oxygen content of molten steel slightly rise during pouring. Therefore, the lining of clay brick can not be used for spraying, especially the use of calcium powder and aluminum powder is not suitable. At present, most of the injection linings are made of high aluminum or magnesium carbon.

可以看出，黏土砖包衬在喷吹过程中侵蚀严重，且有大量的 SiO_2 被还原而使钢中的 Si 含量增加，并影响 Ca 和 Al 的利用率，使钢液在浇注过程中氧含量略有回升。所以，黏土砖的包衬不能用来喷吹，尤其是采用钙系粉剂和铝质粉剂更不适用。目前，喷吹的包衬多采用高铝质或镁碳质。

(2) Indirect deoxidation. After the reduction thin slag is made, the deoxidizer (usually pow-

der deoxidizer) is added on the slag surface to deoxidize the molten steel by reducing the oxygen content in the slag. This deoxidization method is called indirect deoxidization. The theoretical basis of indirect deoxidation is the distribution law, that is, under a certain temperature, the ratio of the activity of oxygen in molten steel to that in slag (FeO) is a constant, which is expressed as:

$$L_0 = a_{(O)}/a_{(FeO)}$$

Where L_0——the distribution coefficient of oxygen.

（2）间接脱氧。还原稀薄渣造好后，将脱氧剂（一般以粉状脱氧剂为主）加在渣面上，通过降低渣中的氧含量来达到钢液的脱氧，这种脱氧方法称为间接脱氧。间接脱氧的理论根据是分配定律，即在一定的温度下，钢液中氧的活度与渣中（FeO）的活度之比是一个常数，表示为：

$$L_0 = a_{(O)}/a_{(FeO)}$$

式中　L_0——氧的分配系数。

When the powder deoxidizer is added to the slag, the content of FeO in the slag is bound to be reduced, and the distribution balance of oxygen between the slag and steel is destroyed. In order to achieve the rebalance, the oxygen in the molten steel will diffuse or transfer to the slag, thus continuously reducing the oxygen content in the slag, the oxygen in the molten steel will be successively removed. Therefore, indirect deoxidation is also called diffusion deoxidation.

将粉状脱氧剂加入渣中，渣中（FeO）的含量势必减少，氧在渣钢间的分配平衡遭到破坏，为了达到重新平衡，钢液中的氧就向渣中扩散或转移，由此不断地降低熔渣中的氧含量，就可使钢液中氧陆续得以脱除。因此，间接脱氧又称为扩散脱氧。

（3）Comprehensive deoxidation. The essence of comprehensive deoxidation is the comprehensive application of direct deoxidation and indirect deoxidation. In the process of operation, strive to overcome their own shortcomings, focus on the advantages to complete the deoxidation task of molten steel. The deoxidization method can not only ensure the quality of steel, but also shorten the reduction time, so it is quite common in production at present.

（3）综合脱氧。综合脱氧的实质就是直接脱氧和间接脱氧的综合应用。在操作过程中，力求克服各自的缺点，集中优点来完成钢液的脱氧任务。该法脱氧既能保证钢的质量，又能缩短还原时间，因此目前在生产上比较常见。

3.4.1.3　Deoxidation of Liquid Steel
3.4.1.3　钢液的脱氧操作

In addition to powder spraying deoxidization, there are many kinds of deoxidization in the furnace. There are two kinds of common deoxidization methods: white slag method and carbide slag method. Their main differences are: white slag does not contain CaC_2, the slag making time is short, and it is suitable for all kinds of steel; carbide slag contains CaC_2, with strong reduction capacity, but the smelting time is long, and it is not used in general steel. In addition, there are deoxidation operations such as neutral slag.

钢液除喷粉脱氧外，炉中脱氧还有多种，比较常见的有白渣法和电石渣法两种，它们的主要区别是：白渣中不含有 CaC_2，造渣时间短，适用于各类钢种；电石渣中含有 CaC_2，还原能力强，但冶炼时间长，在一般钢种上不使用。除此之外，还有中性渣等脱氧操作。

(1) Deoxidation of white slag. After the oxidation slag is removed, the thin slag should be added immediately to reduce the suction and cooling of liquid steel as much as possible. The volume ratio of lime and fluorite in the thin slag is 3.5:1. For the medium capacity furnace, the amount of slag is generally 3% ~ 3.5% of the steel water, and the upper limit can be taken for the small furnace. In order to make the slag melt quickly and form the molten steel covered by the slag, a large power supply and slag pushing and stirring shall be used until the slag with good fluidity is formed. It is better to cause the thin slag at one time to avoid thickening and thinning sometimes in the reduction process. Try to make the slag material ratio and slag quantity accurate and use the current and voltage reasonably.

(1) 白渣脱氧操作。扒净氧化渣后，要立即迅速加入稀薄渣料，尽量减少钢液的吸气与降温。稀薄渣料中石灰和萤石的体积比为3.5:1，对于中等容量的炉子，渣料的加入量一般为钢水量的3% ~ 3.5%，小炉子可取上限值。为使渣料快速熔化形成渣液覆盖钢液，应用较大的功率供电及推渣搅拌，直至形成流动性良好的熔渣。稀薄渣最好一次造成，避免在还原过程中时而调稠、时而调稀，尽量做到造渣材料比及渣量准确，合理地使用电流电压等。

According to the process requirements, before or with the addition of the thin slag, the bulk deoxidizer is added for pre deoxidization. Among them, manganese (including residual manganese content in steel) in steel reaches (close to) the lower specification limit, and silicon reaches 0.10% ~ 0.15%. After the thin slag is formed, the alloy is transferred into the alloy, and then deoxidized powder is added in batches according to the regulations. The batch dosage should be more, and the other batches should be reduced in sequence, with the interval between each batch greater than 6min. About 50% of ferrosilicon powder added as deoxidizer is used for deoxidization or burning loss, and the rest enters the molten steel. Therefore, the total amount of ferrosilicon powder should not be too much, generally 3 ~ 6kg/t steel. Some still use silicon calcium powder or aluminum powder to continue deoxidization. At this time, it is necessary to control the amount of silicon iron powder to prevent silicon in steel from exceeding the specification. After the first batch of ferrosilicon powder is added, 0.3kg/t steel carbon powder shall be added, and other batches shall be added appropriately according to the slag condition. Carbon powder reacts with oxide in slag to form CO gas, which can keep positive pressure in furnace and prevent reduction slag from being oxidized by air. Before each batch of deoxidizing powder is added, the molten pool shall be fully stirred. After adding, the furnace door, electrode hole and charging hole shall be closed tightly to avoid cold air entering the furnace and reducing the reducing atmosphere.

根据工艺要求，薄渣料加入前或随同薄渣料一起加入块状脱氧剂进行预脱氧。其中，使钢中锰（包括钢中残余锰含量）达到（接近）规格下限，硅（质量分数）达到0.10% ~ 0.15%。薄渣形成后调入合金，然后按规程要求分批加入脱氧粉剂；第一批用量应多些，其他各批依次递减，每批间隔大于6min。作为脱氧剂加入的硅铁粉，大约有50%用于脱氧或烧损，余者进入钢液中，因此硅铁粉的总用量不能太多，一般为3 ~ 6kg/t钢。有的还使用硅钙粉或铝粉继续脱氧，这时更要控制硅铁粉的用量，以免钢中硅超出规格。第一批硅铁粉加入后，应加入 0.3kg/t 钢炭粉，其他各批根据渣况适量加入。炭粉和渣中的氧化物反应生成 CO 气体，能使炉内保持正压，进而防止还原渣被空气氧化。脱氧粉剂每批加入前，应对熔池进行充分的搅拌，加完后要紧闭炉门，密封电极孔和加料孔等，避免冷空气进入炉内而降低还原气氛。

The activity of (SiO_2) and (MnO) in high basicity reductive slag is low, and their deoxidization ability can be improved significantly between slag steel interface. In order to make ferrosilicon powder give full play to deoxidation, a proper amount of lime should be added to ferrosilicon powder to increase the local alkalinity of slag. The amount of lime is mainly determined by experience and the fluidity of slag. Silicon carbide powder can also be used instead of ferrosilicon powder. However, the temperature of the molten pool must be higher when SiC powder is added. SiO_2, the deoxidizing product of SiC powder, and SiO_2 in SiC powder can dilute the slag, and the amount of fluorite should be reduced when making the slag. The addition amount of each batch of silicon carbide powder should not be too much, because too much is added, and the steel liquid is easy to be carburized; if the carbon powder slag process is used for deoxidization in the early stage or a large amount of carbon added steel liquid is used when the slag is fully raked, the powder should not be used for deoxidization. In addition, the melting point of silicon carbide is very high (2450~2950℃), which is relatively stable, difficult to melt and decompose, and the silicon content in silicon carbide powder is relatively low, and the yield is also low. Therefore, when using silicon carbide powder for deoxidization, the silicon content in molten steel often cannot meet the specification requirements of steel. In this case, ferrosilicon should be used for silicon adjustment, or ferrosilicon should be added at the same time of deoxidization of silicon carbide powder.

高碱度还原性渣中（SiO_2）和（MnO）的活度较低，在渣钢界面间可显著提高它们的脱氧能力。为使硅铁粉充分发挥脱氧作用，在加入的硅铁粉中应掺入适量的石灰，使渣中局部碱度增高，石灰的掺入量主要凭经验并根据熔渣的流动性而定，也可用碳化硅粉代替硅铁粉。但碳化硅粉加入时，熔池的温度一定要高一些。碳化硅粉的脱氧产物 SiO_2 和碳化硅粉中的 SiO_2 均能稀释熔渣，造稀薄渣时萤石的用量应酌减。碳化硅粉每批的加入量不能太多，因为加入太多，钢液易增碳；如果前期利用炭粉渣工艺脱氧或全扒渣时用炭粉大量增碳的钢液，不宜使用该种粉剂进行脱氧。此外，碳化硅的熔点很高（2450~2950℃），比较稳定，难熔化、分解，且碳化硅粉中的硅含量较低，收得率也低，因此在使用碳化硅粉脱氧时，钢液中的硅含量往往不能满足钢种的规格要求，这时就要利用硅铁调硅，或在碳化硅粉进行脱氧的同时补加硅铁。

In the reduction process, it is necessary to control the temperature and strengthen the mixing to make the temperature and chemical composition uniform and the deoxidized product fully float up, and pay attention to the chemical reaction and change in the molten pool to ensure sufficient alkalinity and proper fluidity. When the slag is thick, fluorite or firebrick shall be added for adjustment, lime shall be added when the slag is thin, if there is a large piece of magnesia floating up, it shall be removed immediately, and the slag shall be replaced when it is serious. In general, the total slag amount in reduction period is 4%~5% of the material weight, the upper limit is taken for small furnace and the lower limit is taken for large furnace, while the alkalinity of white slag should be kept around 3.0.

在还原过程中，要控制好温度和加强搅拌，促使温度、化学成分均匀及脱氧产物充分上浮，并注意熔池内的化学反应与变化，保证有足够的碱度和合适的流动性，渣稠时应加入萤石或火砖块调整，渣稀时应补加石灰，如有大块镁砂浮起应立即扒出，严重时要换渣。一般还原期的总渣量为料重的 4%~5%，小炉子取上限，大炉子取下限，而白渣的碱度应保持在 3.0 左右。

At the beginning of the reduction phase, there are a lot of (FeO) and (MnO) in the slag, which are bright black, and the oxides containing chromium are green. With the deoxidization, the concentration of these oxides decreases and finally turns white. The white slag with good fluidity is active foam, and can stick 2~3mm loose and crisp slag layer on the rake bar. It will quickly break into white sheets when cooled and disperse into white powder over a long time. This is because there is ($2CaO \cdot SiO_2$) in the slag, which is very stable when the temperature is higher than 675℃. When the temperature is lower than 675℃, crystal transformation will occur, from β crystal to γ crystal, resulting in the increase of volume and pulverization. If the content of (SiO_2) in the slag is low and the content of (MnO) is high, it is not easy to break into powder, then the composition of the slag should be adjusted. To evaluate the quality of white slag, not only the degree of slag white, but also the time of slag color keeping in the furnace. The color of white slag is stable, which indicates that the indirect deoxidation of molten steel is good. If the color of slag changes repeatedly, it indicates that the deoxidation is bad. When the deoxidation process is completed, the slag in good condition turns white, and the steel composition can be sampled and analyzed after full stirring.

还原期的熔渣初期因含有大量的（FeO）和（MnO）而呈亮黑色，含有铬的氧化物呈绿色，随着脱氧反应的进行，这些氧化物的浓度不断降低，最后变为白色。流动性良好的白渣，活跃起泡沫，并能在耙杆上均匀粘 2~3mm 松而脆的渣层，冷却时很快破裂成白色片状，时间一长就散成白色粉末。这是因为渣中含有（$2CaO \cdot SiO_2$），在温度高于 675℃时很稳定，当冷却至 675℃以下时，会发生晶体转变，由 β 晶型变为 γ 晶型，致使体积增大而粉化。如果渣中（SiO_2）含量低，而（MnO）含量高时，不易碎成粉末，这时应调整一下熔渣的成分。评定白渣的好坏，不但要看渣白的程度，还要看炉内渣色保持的时间。白渣颜色稳定，说明钢液间接脱氧好，如渣色反复变化，表明脱氧不良。当脱氧工艺完成后，渣况良好渣色变白，经充分搅拌后即可取样分析出钢成分。

There is another way to deoxidize the white slag. When the thin slag is formed, the mixture of carbon powder and ferrosilicon powder (or calcium silicon powder) is used for indirect deoxidization at the same time. This operation is commonly known as the silicon carbon powder slag method. The first batch of deoxidizer is composed of 1~3kg/t steel carbon powder and 1~3kg/t steel ferrosilicon powder (or calcium silicon powder), which is kept for 10~15min after being added into the furnace, and then added into ferrosilicon powder in batches after the slag turns white. The practice shows that the deoxidization effect of this method is also very good, and it has been widely used in production.

白渣脱氧操作还有另外一种,就是当稀薄渣形成后,用炭粉和硅铁粉(或硅钙粉)混合物同时进行间接脱氧,这种操作俗称硅炭粉渣法。硅炭粉渣的第一批脱氧剂是由1~3kg/t钢炭粉和1~3kg/t钢硅铁粉(或硅钙粉)组成,加入炉中后保持10~15min,渣变白后再分批加入硅铁粉等继续脱氧。实践证明,该法脱氧效果也很好,并在生产上获得了较为广泛的应用。

(2) Deoxidation operation of carbide slag. After cleaning the oxidation slag, add pre deoxidizer according to the process requirements, and then make thin slag immediately. The slag condition is slightly thicker, and then add carbon powder 1.5~3.0kg/t steel. In order to accelerate the formation of calcium carbide slag, the furnace door should be closed after the addition of carbon powder, the electrode hole and charging hole should be sealed, and a larger power should be used. The carbon powder reacts with calcium oxide in the high temperature area to generate calcium carbide:

$$3C + (CaO) = (CaC_2) + \{CO\}$$

(2) 电石渣脱氧操作。扒净氧化渣后,根据工艺要求加入预脱氧剂,然后立即造稀薄渣,渣况稍稀一点,再加入炭粉1.5~3.0kg/t钢。为了加速电石渣的形成,炭粉加入后应紧闭炉门,封好电极孔和加料孔,并使用较大的功率,炭粉在高温区与氧化钙发生反应生成碳化钙:

$$3C + (CaO) = (CaC_2) + \{CO\}$$

This is a strong endothermic reaction, so a large current and voltage must be used to ensure a high temperature in the furnace. When there is thick black smoke or flame with black smoke coming out of the gap of the furnace, it indicates that carbide slag has been formed. In order to reduce the time of forming carbide slag and the reduction period, small pieces of 3~5kg/t carbide steel can also be directly added to the steel surface or with the formation of thin slag or slag, and then a small amount of carbon powder can be added to keep the positive pressure in the furnace for deoxidization, and the same effect can be obtained.

这是个强吸热反应,因此必须使用较大的电流与电压,以保证炉内具有很高的温度。当有浓浓的黑烟或带黑烟的火焰从炉子的缝隙冒出时,标志着电石渣已形成。为了减少形成电石渣的时间,缩短还原期,也可往钢液面上或随稀薄渣料或稀薄渣形成后直接加入小块电石3~5kg/t钢,然后再调入少量炭粉使炉内保持正压进行脱氧,同样能得到相同的效果。

Calcium carbide can be dissolved and diffused in slag. The deoxidation reaction is as follows:

$$3(FeO) + (CaC_2) = 3[Fe] + (CaO) + 2\{CO\}$$
$$3(MnO) + (CaC_2) = 3[Mn] + (CaO) + 2\{CO\}$$

碳化钙在渣中既能溶解又能扩散,脱氧反应如下:

$$3(FeO) + (CaC_2) = 3[Fe] + (CaO) + 2\{CO\}$$
$$3(MnO) + (CaC_2) = 3[Mn] + (CaO) + 2\{CO\}$$

When the calcium carbide is deoxidized, the carbon powder used to keep the positive pressure can also reduce (FeO) and (MnO) in the slag in the lower temperature region.

在碳化钙进行脱氧的同时,为了保持正压而使用的炭粉在较低的温度区域内也能还原渣中的（FeO）和（MnO）。

Carbide slag is a kind of reducing slag with high basicity, and its deoxidizing ability is stronger than that of white slag. In order to fully deoxidize the molten steel, the carbide slag should be kept for 20~30min. In addition to the poor temperature control, the molten steel is easy to increase silicon, about 0.05%~0.15%/h. This is due to the reaction between CaC_2 and (SiO_2) in the slag. The reaction formula is as follows:

$$3(SiO_2) + 2(CaC_2) = 3[Si] + 2(CaO) + 4\{CO\}$$

电石渣是高碱性的还原熔渣，脱氧能力比白渣强。为使钢液充分脱氧，电石渣应保持20~30min。在电石渣下操作，除温度不好控制外，钢液还易增硅，增硅0.05%~0.15%/h，这是由于CaC_2与渣中（SiO_2）发生反应的结果，反应式如下：

$$3(SiO_2) + 2(CaC_2) = 3[Si] + 2(CaO) + 4\{CO\}$$

In addition, it is easy to increase the carbon content of molten steel by 0.05%~0.10%/h. If the carbide slag is used for tapping, the free carbon in the slag will also increase the carbon content of the finished steel.

此外，还易使钢液增碳，增碳0.05%~0.10%/h。如果采用电石渣出钢，渣中的游离碳也会使钢的成品碳增高。

Calcium carbide slag can be divided into weak carbide slag [$w(CaC_2) = 1\% \sim 1.5\%$] and strong carbide slag [$w(CaC_2) = 2\% \sim 4\%$] due to the different content of calcium carbide. It is not easy to distinguish the two from each other, except that the weak carbide slag is gray black and the strong carbide slag is pitch-black, but they are obviously different from the oxide slag. The carbide slag has no luster, while the oxide slag is bright black. In addition, the carbide slag sometimes has white stripes, which will be crushed when it is cooled in the air. When it is put into the water, acetylene (C_2H_2) gas can be released, and it has a strong pungent smell. The reaction formula is as follows:

$$CaC_2 + 2H_2O = C_2H_{2(气)} + Ca(OH)_2$$

电石渣因碳化钙含量不同分为弱电石渣 [$w(CaC_2) = 1\% \sim 1.5\%$] 和强电石渣 [$w(CaC_2) = 2\% \sim 4\%$]。两者从现象上区分不太容易，只不过弱电石渣显灰黑色，强电石渣呈乌黑色，但它们和氧化渣有明显的区别，电石渣无光泽，而氧化渣呈亮黑色。另外，电石渣有时带有白色条纹，在空气中冷却会粉碎，放入水中能放出乙炔（C_2H_2）气体，并有强烈的刺激味，反应式如下：

$$CaC_2 + 2H_2O = C_2H_{2(g)} + Ca(OH)_2$$

Because the melting point of CaC_2 is high, the viscosity of carbide slag is high and it is well wetted with molten steel. From tapping to final casting, it is easy to increase inclusions in steel because slag steel is not easy to separate, so it is generally required that carbide slag be turned into white slag before tapping.

因为CaC_2的熔点高，所以电石渣黏度大，并与钢液润湿好。从出钢到最终浇注，由于渣钢不易分离而易使钢中夹杂增加，因此一般均要求电石渣变成白渣后方能出钢。

The deoxidation of carbide slag can be judged by the change of slag color. At first, the content of (FeO) and (MnO) in the slag is high, and the slag is bright black. With the decrease of oxide in the slag, it becomes brown black. When the deoxidation is good, the slag surface has no luster or white stripe, and the slag color turns white completely, indicating that deoxidation is good. After the carbide slag turns white, ferrosilicon powder or calcium silicon powder should be added in batches to continue deoxidization. Each batch of steel should be added with the amount of 1~1.5kg/t, and one batch should be added every 6~7min. Before adding, the white slag should be stirred and kept until tapping.

电石渣的脱氧情况可根据熔渣颜色的变化来判断。最初，熔渣中（FeO）和（MnO）含量较高，渣呈亮黑色，随着渣中氧化物的减少，变成棕黑色。当脱氧比较好时，熔渣表面无光泽或带有白色条纹，渣色完全变白，说明脱氧良好。电石渣变白后，一般还要分批加入硅铁粉或硅钙粉继续脱氧，每批加入量为1~1.5kg/t 钢，每隔6~7min 加一批，加前要搅拌，并使白渣保持到出钢。

In order to make the carbide slag white in time, in addition to the appropriate amount of carbon powder, it is necessary to control the current and voltage. If it is difficult to turn white, it means that there is too much free carbon and calcium carbide in the slag. In this case, it is necessary to dilute or oxidize the free carbon and calcium carbide in the slag: add lime and fluorite to the molten pool for dilution; open the furnace door and charging hole, let the air pass through the furnace, and oxidize the free carbon and calcium carbide in the slag; if necessary, blow a small amount of oxygen into the slag surface or some of the slags are removed and replaced with new ones.

为使电石渣及时变白，除炭粉的用量要合适外，还要控制好电流和电压。如果一旦难以变白，说明渣中游离碳和碳化钙过多，这时应采取稀释或氧化掉渣中游离碳及碳化钙的办法：可向熔池中加入石灰、萤石进行稀释；也可打开炉门及加料孔，让空气穿膛而过，将渣中的游离碳和碳化钙氧化掉；迫不得已时，还可向渣面上吹入少量的氧或采取部分扒渣补换新渣等办法进行处理。

(3) Deoxidation of neutral slag. After the oxidation slag is removed, pre deoxidizer is added according to the technological requirements of steel grades, and then neutral thin slag is produced. The slag of neutral slag is mainly composed of firebrick mixed with lime and fluorite, so that the content of Cao and MgO in the slag is almost the same as that of SiO_2, and MgO is obtained from thin slag and high temperature erosion of furnace lining for a long time. Because the main components of the slag are MgO and SiO_2, it is also called $MgO-SiO_2$ neutral slag.

(3) 中性渣脱氧操作。扒净氧化渣后，根据钢种的工艺要求加入预脱氧剂，然后造中性稀薄渣。中性渣的渣料主要由火砖块掺入适量的石灰和萤石组成，使渣中的（CaO）与（MgO）的含量之和几乎同（SiO_2）的含量相当，（MgO）是由稀薄渣及高温长时间侵蚀炉衬而得。由于这种渣的主要成分为MgO 和 SiO_2，所以又称为 $MgO-SiO_2$ 中性渣。

After the formation of neutral thin slag, it is necessary to add deoxidizing powder in batches immediately for indirect deoxidization, and the quantity used shall be determined according to the silicon content in the molten steel and the quantity of basic oxide in the slag. The neutral slag can stabilize the carbon in the molten steel, and the resistance is large, which is conducive to the heating and heating of the molten steel. In addition, the surface tension of the slag is large, which is beneficial to the stability of arc light. However, this kind of slag has serious erosion on the basic furnace lining, so the slag amount should not be too large in the slag making process, generally about 1% ~2% of the steel water amount. In addition, this kind of slag has no desulfuration ability, so it has strict requirements on furnace charge. It is rarely used except for smelting sulfur-containing free cutting steel or some stainless steel.

中性稀薄渣形成后,要立即分批加入脱氧粉剂进行间接脱氧,使用数量应参照钢液中的硅含量和熔渣中的碱性氧化物的数量而定。中性渣能稳定钢液中的碳,且电阻大,有利于钢液的加热和升温。另外,这种渣表面张力大,有利于弧光稳定。但该种渣对碱性炉衬侵蚀严重,所以造渣过程中,渣量不宜过大,一般为钢水量的1%~2%。此外,这种渣没有脱硫能力,因此对炉料要求比较严格,除冶炼含硫易切削钢或某些不锈钢外,已很少采用。

3.4.1.4 Inspection of Deoxidizing Effect of Molten Steel
3.4.1.4 钢液脱氧效果的检验

At present, although there is no unified standard for the test of deoxidizing effect of molten steel, for the molten steel with simple deoxidizing process or using weak deoxidizer to deoxidize and pour killed steel, the test of deoxidizing effect is still needed after deoxidizing operation. If bad deoxidation is found, active and effective measures should be taken immediately.

目前,钢液脱氧效果的检验虽然尚未制定统一的标准,但电炉炼钢工对于脱氧工艺简单或使用弱脱氧剂脱氧并用于浇注镇静钢的钢液,在脱氧操作结束后,一般仍要进行脱氧效果的检验。如果发现脱氧不良,就要立即采取积极有效的措施加以处理。

There are many methods to test the deoxidation effect of liquid steel, and the following are more common:

(1) Experience judgment. The experienced steelmaker can roughly judge the deoxidation quality of molten steel from the slag color: if the slag is white and can keep stable for a long time in the furnace, and the fracture surface is gray and can be powdered and broken in the air, it indicates that the content of FeO is low, then it can be judged that the deoxidation of molten steel is good.

钢液脱氧效果检验的方法较多,较为常见的有以下几种:

(1) 经验判断法。有经验的电炉炼钢工从炉渣的颜色上可大概地判断出钢液脱氧的好坏:如果渣白且在炉中能保持长时间的稳定,而断口呈灰白并能在空气中粉化、碎裂,表明渣中(FeO)的含量较低,这时可判断为钢液脱氧良好。

(2) Deoxidation cup observation method. The observation method of deoxidation cup is the original method to judge the deoxidation of molten steel. After deoxidization, the liquid steel is gently injected into a clean, dry and round high cylinder cup. After solidification, the surface is calm or shrinks in varying degrees, which indicates that the liquid steel is deoxidized well [see Figure 3-2(a)]; if a bunch of sparks emerges during solidification or does not shrink after solidification, instead, there is rising and protrusion, which may be a sign of poor deoxidation [see Figure 3-2(b)].

（2）脱氧杯观察法。脱氧杯观察法是判断钢液脱氧好坏比较原始的方法。脱氧结束后，将钢液轻轻地注入清洁、干燥、圆形的高筒杯内，凝固后表面平静或有不同程度的收缩，说明钢液脱氧良好［见图3-2(a)］；如凝固过程中冒出一束束火花或在凝固后不但不收缩，反而有上涨、突起现象，这可能是脱氧不好的标志［见图3-2(b)］。

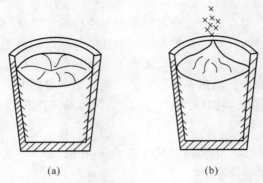

Figure 3-2　Observation of deoxidation with deoxidation cup
(a) well deoxidized sample; (b) poorly deoxidized sample
图3-2　利用脱氧杯观察脱氧
(a) 脱氧良好的试样；(b) 脱氧不良的试样

Sometimes the deoxidation of molten steel is good, but it will rise because of the high hydrogen content in the steel. The identification method is as follows: take out the molten steel, insert a small amount of aluminum for forced deoxidation, and then inject the molten steel into the deoxidation cup gently. If it does not rise, it indicates that the deoxidation is bad; if it still rises, it indicates that the hydrogen content in the steel is high. In addition, if the fully deoxidized liquid steel is injected into a wet and unclean deoxidizing cup, or the amount of liquid steel injected is too small or too strong, it is easy to cause rise. Therefore, when using the deoxidizing cup to test the deoxidizing effect, the deoxidizing cup must be clean and dry, and the liquid steel should be injected gently, and the amount of molten steel should be at least more than 60% of the height of the deoxidizing cup. For high carbon steel with sufficient deoxidation but temperature higher than 1600℃ and carbon content higher than 1%, such as T12A, when the molten steel solidifies in the deoxidizing cup, a bunch of sparks will often appear, or there will rise and form a hard cover, giving a false impression of poor deoxidation. If the hard cover is poked open, it is found that the following shrinkage is still good, indicating that the deoxidation of the liquid steel is still good. The appearance of this illusion may be due to the decomposition of the cementite Fe_3C in steel when the temperature is higher than 1600℃, and the secondary reaction between the decomposed free carbon and oxygen occurs with the decrease of temperature. For high tungsten steel, such as 3Cr2W8V, sometimes the above similar phenomenon occurs. The reason may be that the molten steel of high tungsten steel will continue to react with oxygen as the temperature decreases during solidification and crystallization, or it may be the cementite containing tungsten, such as $(Fe,W)_3C$ or other complex and low melting point carbides, such as CrC_6, etc., which will decompose at high temperature, it is caused by the reaction of carbon and oxygen. Generally, the higher the temperature is, the faster the injection speed is, and the more significant the above-mentioned artifacts are.

有时钢液脱氧良好，但因钢中的氢含量较高，也会引起上涨，鉴别方法如下：将钢液取出，插入少量的铝进行强制脱氧，然后再将钢液轻轻注入脱氧杯内，如不上涨，说明脱氧不良；如仍然上涨，说明钢中的氢含量较高。此外，经过充分脱氧的钢液，如果注入潮湿、不干净的脱氧杯中，或注入的钢水量太少或太猛，也容易引起上涨。因此，利用脱氧杯检验脱氧效果时，脱氧杯必须清洁、干燥，并将钢液轻轻地注入，钢水量最少也应超过脱氧杯高度的60%以上。对于经过充分脱氧，但温度高于1600℃、碳含量大于1%的高碳钢，如T12A等，当钢液在脱氧杯内凝固时，也往往冒出一束束火花，或有上涨并结成一层硬盖，造成一种似乎脱氧不良的假象。如将硬盖捅开，发现下面的收缩照样良好，表明该钢液脱氧仍然较好。这种假象的出现，可能是当温度高于1600℃时，钢中的渗碳体Fe_3C发生分解，随着温度的降低，分解出来的游离碳与氧发生二次反应的结果。对于高钨钢，如3Cr2W8V等，有时也出现上述类似现象，原因可能是高钨钢的钢液，在凝固结晶过程中，随着温度的降低，首先析出的碳与氧继续发生反应，也有可能是含钨的渗碳体，如$(Fe,W)_3C$或其他的复杂、低熔点的碳化物，如CrC_6等较多，并在高温下分解而产生的游离碳与氧再次发生反应而造成的。一般温度越高，注速越快，上述假象越显著。

(3) Chemical analysis. The content of FeO in reducing slag can be determined quickly by chemical analysis. If the content of FeO in slag is less than 0.5%, the deoxidation of molten steel is better.

(3) 化学分析法。化学分析能够快速测出还原渣中（FeO）的含量。如果渣中$w(FeO) < 0.5\%$，表明钢液的脱氧较好。

(4) Instrument measurement. At present, the oxygen content in molten steel can be measured directly by the instrument. The instruments used are fast oxygen and carbon determinator, electronic potentiometer, concentration cell oxygen determinator, temperature and oxygen determinator. These meters are easy to use, fast and accurate. Some of them can know the oxygen content in steel in a few seconds. Therefore, they have been widely used in many steel-making units.

(4) 仪表测量法。目前，通过仪表可直接测出钢液中的氧含量。所用的仪表有快速定氧定碳仪、电子电位差计、浓差电池定氧仪、测温定氧仪等。这些仪表使用方便、快速、准确，有的只需几秒就可确切知道钢中的氧含量，因此，已在许多炼钢企业得到普遍的应用。

3.4.1.5　Final deoxidation of Molten Steel
3.4.1.5　钢液的终脱氧

In order to further reduce the oxygen content in the steel, according to the process requirements, the elements with stronger deoxidizing ability can also be added into the ladle or in the pouring process before tapping or to the tapping tank and tapping flow, that is, the last direct deoxidizing of the liquid steel before solidification, which is called the final deoxidizing of the liquid steel. After the final deoxidization of the molten steel, the ideal crystal structure can be obtained after solidification, and various properties are also improved.

为了进一步降低钢中氧含量，根据工艺要求，在出钢前向出钢槽、出钢流中，也可在钢包中或在浇注过程中加入脱氧能力更强的元素，即在凝固前对钢液进行最后一次直接脱氧，称为钢液的终脱氧。经过终脱氧的钢液，凝固后可获得理想的结晶组织，各种性能也得到了提高。

Due to the different amount, type or operation of final deoxidizer, the influence on the content, shape, size and distribution of inclusions in steel is also different. The commonly used final deoxidizers are aluminum, titanium, calcium silicon or cerium iron.

由于终脱氧剂的用量、种类或操作不同，对钢中夹杂物的含量、形状、大小与分布的影响也不同。常用的终脱氧剂有铝、钛、硅钙或铈铁等。

The final deoxidization of aluminum is mostly used to insert into steel 2~3 minutes before tapping. The general amount of deoxidization is: low carbon steel is 1kg/t steel, medium carbon steel is 0.8kg/t steel, high carbon steel and high silicon steel are 0.5kg/t steel, and tool steel is 0.5kg/t steel. The present wire feeder can save aluminum consumption by feeding aluminum core into steel. Using aluminum as the final deoxidizer can control the secondary oxidation of molten steel and reduce the formation of bubbles in ingot, at the same time, it can prevent the formation of Fe_4N and avoid the aging of steel. In addition, the combination ability of aluminum and nitrogen is relatively large. In the case of good deoxidization, aluminum nitride can be generated from the final deoxidized aluminum and nitrogen in steel, and distributed in the steel in a highly dispersed state. When the steel is crystallized, the grains can be refined as non spontaneous cores. For steel grades with high grain size requirements, it is ideal to keep the residual total aluminum content in the range of 0.02%~0.05%. Before the final deoxidized aluminum is added, it is generally cast into a uniform block with a single weight of no more than 3kg, and then inserted into the molten pool according to the amount of consumption. Before inserting, it is better to put it on the furnace door and bake it. When inserting, the aluminum block should quickly pass through the slag layer and enter into the steel. After a while, the rake bar should be moved. After the aluminum is blocked, the rake bar should be taken out. In order to expand the final deoxidized range of aluminum in the molten pool, stirring should be carried out after inserting. The final deoxidization with aluminum block in ladle will affect the normal pouring of steel due to the low density of aluminum, easy floating, or the flocculent distribution of deoxidized product Al_2O_3 in steel, hanging or blocking the channel on the inner wall of ladle nozzle. At present, a spray device has been developed to spray the prepared aluminum shot into the deep part of the molten steel, so that it can be evenly distributed in the steel, so as to not only meet the requirements of smelting process, but also improve the utilization rate of final deoxidized aluminum. In the pouring process, the final deoxidization of steel can also be carried out by adding aluminum shot or aluminum wire into the central injection pipe. Titanium oxides and carbides can also become the non spontaneous core of molten steel during crystallization, so they can also refine the grains. TiO_2, the deoxidizing product of titanium, can reduce the melting point of manganese or iron silicate, and promote the aggregation and floatation of silicate inclusions. As a result, the content of such inclusions in steel is significantly reduced. In addition, titanium is also the strongest element to form tin. Although it can reduce the harm of nitrogen to the steel quality, there will always be some remaining in the steel to form angular inclusions. Titanium iron is generally used for final deoxidation, with small amount and no burning loss. Before adding, the aluminum 1kg/t steel should be inserted to ensure the final deoxidization effect of titanium. When adding, it is better to push away the slag surface to make it directly contact with the molten steel. It can also be added into the tapping tank or tapping flow. As the final deoxidizing agent, silicalcium can reduce Al_2O_3 chain inclusions and increase spherical inclusions in steel, control the secondary oxidation, desulfurization and degassing of liquid steel, improve the fluidity of liquid steel, the surface quality of ingot and the impact toughness of steel. The amount of final deoxidized calcium silicate block is generally no more than 1kg/t steel, which is directly put into the steel flow or ladle when tapping. Cerium is good for deoxidization, desulfuration, degassing and grain refinement of steel. It can also reduce the secondary oxidation of liquid steel and improve the mechanical properties of steel. The addition amount of cerium iron is generally 0.3 ~ 0.5kg/t steel, which is mostly directly put into the steel flow or ladle. If it is used in the furnace, it should be wrapped with iron sheet and inserted into the molten steel.

铝的终脱氧多用在出钢前 2~3min 插入钢中，一般用量：低碳钢为 1kg/t 钢，中碳钢为 0.8kg/t 钢，高碳钢和高硅钢为 0.5kg/t 钢，而工具钢为 0.5kg/t 钢。目前出现的喂丝机将制成的铝芯喂入钢中，可以节省铝的用量。用铝作为终脱氧剂可控制钢液的二次氧化及减少钢锭产生气泡，同时又能防止 Fe_4N 生成，避免钢产生老化。此外，铝和氮的结合能力较强，在脱氧良好的情况下，终脱氧铝和钢中的氮可生成氮化铝，并以高度弥散状态分布在钢中，在钢液结晶时可作为非自发核心而细化晶粒。对于晶粒度要求较高的钢种，残余全铝量保持在 0.02%~0.05% 范围内较为理想。终脱氧铝加入前，一般先铸成单重不大于 3kg 的均匀块体，然后根据用量的多少插入熔池中，插前最好先放在炉门上烘烤一下，插时要使铝块迅速穿过渣层进入钢中，稍化一会再活动耙杆，待铝块化后，拿出耙杆，为了扩大铝在熔池中的终脱氧范围，插后应进行搅拌。在钢包中用铝块进行终脱氧，会因铝的密度小，易飘浮，或脱氧产物 Al_2O_3 在钢中呈团絮状分布，在钢包水口的内壁上悬挂或堵塞汤道而影响钢的正常浇注。目前，已研制出一种喷射装置，将预先制备的铝丸喷射到钢液的深部，使其在钢中分布均匀，从而既满足冶炼工艺的要求，又提高终脱氧铝的利用率。在浇注过程中，通过中注管添加铝丸或铝丝也可对钢进行终脱氧。钛的氧化物和碳化物也能成为钢液结晶时的非自发核心，因此也有细化晶粒的作用。钛的脱氧产物 TiO_2 对锰或铁硅酸盐有降低熔点的作用，可以促使硅酸盐夹杂物聚集上浮，结果使钢中的这类夹杂物含量显著减少。此外，钛也是形成 TiN 最强的元素，虽然可减轻氮对钢质量的危害程度，但总会有一部分残留在钢中而形成带棱角的夹杂物。终脱氧一般使用钛铁，用量不大，不计烧损。加入前应插铝 1kg/t 钢，以确保钛的终脱氧效果，加时最好推开渣面，使之与钢液直接接触，也可在出钢槽或出钢流中加入。作为终脱氧的硅钙能减少钢中 Al_2O_3 链状夹杂物，增加球状夹杂物，还能控制钢液的二次氧化、脱硫与除气，改善钢液的流动性和钢锭的表面质量及提高钢的冲击韧性等。终脱氧硅钙块的用量一般不大于 1kg/t 钢，出钢时直接投入钢流或钢包中。铈对钢的脱氧脱硫及除气与细化晶粒均有好处，还能减少钢液的二次氧化并能改善钢的力学性能等。铈铁的加入量一般为0.3~0.5kg/t钢，多直接投入钢流或钢包中，如在炉中使用，要用铁皮包好插入钢液中。

3.4.1.6 Secondary Oxidation and Control of Molten Steel
3.4.1.6 钢液的二次氧化与控制

In the process of tapping and pouring, some elements in the steel may react with oxygen or nitrogen in the air to form secondary oxides and nitrides due to the exposed steel and direct contact with air. In addition, with the decrease of temperature after tapping, the solubility of [O] in steel also decreases, while the reaction ability of these elements with [O] increases, so that oxides will continue to form under new conditions. Similar phenomena are called secondary oxidation of molten steel. The secondary oxidation of molten steel increases the total amount of inclusions in the steel, which seriously affects the properties of the finished steel. Therefore, it should be prevented and avoided in the production process. A good final deoxidation operation can further reduce the oxygen in the steel, thus greatly reducing the degree of secondary oxidation in the molten steel, especially with aluminum. In addition, the second oxidation of liquid steel can also be reduced by shortening the tapping time as much as possible and adopting large orifice spraying and slag steel mixing.

在出钢和浇注过程中，脱氧良好的钢液，由于钢液的裸露并与空气直接接触，钢中某些元素有可能与空气中的氧或氮发生反应，生成二次氧化物及氮化物。此外，出钢后随着温度的降低，[O]在钢中的溶解度也降低，而这些元素与[O]的反应能力却增加，这样在新的条件下又要继续生成氧化物。类似这些现象统称为钢液的二次氧化。钢液的二次氧化使钢中夹杂物的总量明显增加，严重地影响成品钢的各种性能，因此，在生产过程中应尽量防止与避免。良好的终脱氧操作能将钢中的氧进一步降低，从而可使钢液发生二次氧化的程度大大减少，尤其是用铝的终脱氧作用更大。此外，尽量缩短出钢时间，并采取大口喷吐、渣钢混出等，也能减少钢液的二次氧化。

At present, the more popular vacuum smelting or ladle powder spraying is deoxidization after tapping, which can reduce the oxygen content in steel to a very low level, and basically avoid or eliminate the secondary oxidation of molten steel. In addition, inert gas, liquid slag or graphite slag protection pouring and vacuum pouring are also powerful measures to control the secondary oxidation of molten steel.

目前，比较盛行的真空冶炼或钢包喷粉是在出钢后进行脱氧，能使钢中的氧含量降到很低的水平，基本上可避免或消除钢液的二次氧化。除此之外，在浇注过程中，采用惰性气体、液渣或石墨渣保护浇注及真空浇注等，也都是控制钢液二次氧化的有力措施。

3.4.2 Desulfurization in Reduction Period
3.4.2 还原期的脱硫

The desulfuration task of EAF steelmaking in reduction period is mainly completed in reduction period or by refining the molten steel outside the furnace. Desulfuration is one of the important contents of reduction refining.

电炉炼钢的脱硫任务主要是在还原期或利用钢液的炉外精炼来完成的,脱硫是还原精炼的重要内容之一。

Desulfuration can be achieved in every smelting stage of basic electric furnace steelmaking. Generally, L_S in melting period can reach 3~4, in oxidation period 8~10, and in reduction period 50~80. It can be seen that most of the desulfurization tasks are carried out in the reduction period. The following describes the strengthening measures for desulfurization operation in reduction period and tapping process.

碱性电炉炼钢的各个冶炼阶段都能脱硫。一般熔化期的 L_S 可达 3~4,氧化期的 L_S 可达 8~10,而还原期的 L_S 可达 50~80。由此可见,大部分的脱硫任务是在还原期进行的。以下介绍还原期和出钢过程脱硫操作的强化措施。

3.4.2.1 Strengthening Measures of Desulfurization Operation in Reduction Period
3.4.2.1 还原期脱硫操作的强化措施

Strengthening measures for desulfurization operation in reduction period include:
还原期脱硫操作的强化措施有:

(1) Increase the temperature of the whole slagging. Desulfurization is an endothermic reaction, so it is beneficial to increase the temperature of the whole slag removal. In addition, increasing the temperature of full slagging can make the formation of thin slags faster and improve the dynamic conditions of desulfurization.

(1) 提高全扒渣温度。脱硫是吸热反应,提高全扒渣温度有利于反应的进行。此外,提高全扒渣温度还能使稀薄渣形成速度快,以及改善脱硫的动力学条件。

(2) Increase the basicity of reducing slag. When the amount of slag is suitable and the fluidity is good, the basicity R of slag should be kept between 2.5 and 3.5, and in this range, the basicity should be increased as much as possible to make the desulfurization reaction proceed in a favorable direction.

(2) 提高还原渣的碱度。在渣量合适,流动性良好的情况下,渣的碱度 R 应保持在 2.5~3.5,并在这个范围内尽量提高碱度,使脱硫反应朝有利的方向进行。

(3) Strengthen direct pre deoxidation or adopt compulsory deoxidation process. Under the direct action of the pre deoxidizer, the oxygen content in the steel can be reduced rapidly. The lower the oxygen content is, the better the oxygen content is. For example, when the basicity $R = 2.8$ ~ 3.2, $w(FeO) < 0.5\%$, the desulfurization amount of molten steel can reach more than 40%, when $w(FeO) = 0.6\% \sim 1.0\%$, the desulfurization amount of molten steel is about 30%, but when $w(FeO) > 1.0\%$, the desulfurization amount of molten steel is very low. In addition, if the deoxidization process such as carbide slag is used in the reduction period of basic EAF steelmaking, it is also conducive to the desulfurization reaction.

(3) 加强直接预脱氧或采用强制性的脱氧工艺。在预脱氧剂的直接作用下,迅速降低钢中的氧含量,越低越好,可使钢中的硫脱除 20%~30%。例如当碱度 $R = 2.8 \sim 3.2$、

$w(\text{FeO}) < 0.5\%$ 时，钢液的脱硫量可达 40% 以上，$w(\text{FeO}) = 0.6\% \sim 1.0\%$ 时，钢液的脱硫量为 30% 左右，而当 $w(\text{FeO}) > 1.0\%$ 时，钢液的脱硫量却很低。除此之外，碱性电炉炼钢的还原期如果采用电石渣等强制性的脱氧工艺，也有利于脱硫反应的进行。

(4) Strengthen the mixing of molten pool. Strengthening molten pool stirring is an important measure to strengthen desulfurization, especially in the middle and later stages of reduction, when the sulfur content in the molten steel is low, the desulfurization reaction often fails to reach equilibrium, and the diffusion of sulfur is strengthened by stirring to improve the dynamic conditions of desulfurization.

(4) 加强熔池的搅拌。加强熔池搅拌是强化脱硫的一项重要措施，特别是在还原的中后期，当钢液中的硫含量较低时，脱硫反应经常达不到平衡，通过搅拌强化硫的扩散，以改善脱硫的动力学条件。

(5) By adjusting the current and voltage, the desulfurization speed can be improved. In order to obtain the molten steel with low sulfur content, the method of low voltage and high current can be used for desulfurization at the end of reduction. Because the arc light of low voltage and high current is short, the slag can be blown to activate the reaction area, which is conducive to the desulfurization reaction.

(5) 通过调整电流电压，改善脱硫速度。为了获得硫含量很低的钢液，在还原末期可采用低电压、大电流的办法进行脱硫。因低电压、大电流的弧光短，能吹动熔渣而活跃反应区域，进而有利于脱硫反应的进行。

(6) Slag change operation. Under the condition of difficult desulfurization, in order to obtain liquid steel with low sulfur content, partial slagging or complete slagging can be adopted at the end of reduction, and then lime and fluorite can be added to make new slags according to the appropriate ratio, so as to expand the reaction area of slag steel and increase the slag amount, and finally complete the desulfurization task of liquid steel.

(6) 换渣操作。在脱硫困难的条件下，为了获得硫含量很低的钢液，在还原末期可采取部分扒渣或完全扒渣，然后按合适的配比加入石灰和萤石补造新渣，以此来达到扩大渣钢反应界面面积和增大渣量，最终也能较好地完成钢液的脱硫任务。

3.4.2.2 Strengthening Measures for Desulfurization Operation in Tapping Process
3.4.2.2 出钢过程脱硫操作的强化措施

Generally speaking, the ratio of sulfur content in molten steel before tapping to that after tapping is called tapping desulfurization efficiency. With the symbol η_S. Means, then:

$$\eta_\text{S} = \frac{[\text{S}]_\text{before tapping}}{[\text{S}]_\text{after tapping}}$$

通常将出钢前钢液中硫含量与出钢后钢液中硫含量之比，称为出钢脱硫效率。以符号 η_S 表示，则：

$$\eta_\text{S} = \frac{[\text{S}]_\text{出钢前}}{[\text{S}]_\text{出钢后}}$$

There are four strengthening measures for desulfurization operation in tapping process:

出钢过程脱硫操作的强化措施大体上有以下 4 项:

(1) Keep the normal color of reducing slag. The higher the average (FeO) content of reduction slag before and after tapping is, the higher the desulfurization efficiency η_S is. The reduction slag is yellow before tapping, which indicates that the content of (FeO) and (MnO) in the slag is high, which is harmful to the desulfurization of tapping process. Before tapping, the reduction slag is black or gray black, which indicates that there is more free carbon in the slag or carbide slag. If tapping under this kind of slag, it is good for desulfuration, but because the slag is well wetted with the molten steel, the slag steel is not easy to separate after tapping, and it also affects the floating of inclusions, which is not what we hope. The content of Cr_2O_3 in green reduction slag is higher, which has a little good effect on desulfurization. However, the content of FeO in bright black oxide slag is high, which can improve the desulfurization efficiency η_S extremely unfavorable. Therefore, it is an important measure to keep the normal color of reducing slag (white slag or flower white slag) to strengthen desulfurization operation in tapping process.

(1) 保持还原渣的正常颜色。出钢前后还原渣中平均 (FeO) 含量降低越大, 脱硫效率 η_S 提高得越大。出钢前还原渣为黄色, 表明渣中 (FeO) 及 (MnO) 的含量较高, 对出钢过程的脱硫不利。出钢前还原渣呈黑色或灰黑色, 表明渣中游离碳或是电石渣较多, 如果在该种渣下出钢, 对脱硫有好处, 但因熔渣与钢液润湿较好, 出钢后渣钢不易分离, 且又影响夹杂物的上浮, 这也不是我们所希望的。绿色的还原渣中含 Cr_2O_3 较高, 对脱硫略有好的影响。而亮黑色的氧化渣中 (FeO) 的含量高, 对提高出钢脱硫效率 η_S 极为不利。所以, 保持还原渣的正常颜色 (白渣或花白渣) 是出钢过程强化脱硫操作的一项重要措施。

(2) Keep the normal state of reducing slag. This mainly refers to the basicity and fluidity of slag, that is to say, before tapping, the reducing slag is required to have proper basicity and fluidity. Too high and too low basicity is not good for desulfuration in tapping process. When $R = 3.5 \sim 4.2$, the desulfurization efficiency η_S is the highest. However, the lower the original sulfur content $[S]_0$ is, the lower the desulfurization efficiency η_S is. The relationship between basicity and fluidity of slag is very close. Too low basicity affects the desulfuration process, too high basicity destroys the fluidity of slag and hinders the diffusion and transfer of sulfur in slag. Therefore, to keep the proper basicity and good fluidity of reducing slag is another measure to strengthen desulfurization operation in tapping process.

(2) 保持还原渣的正常状态。这里主要是指熔渣的碱度与流动性, 即出钢前要求还原渣要有合适的碱度和合适的流动性。碱度过高过低对出钢过程的脱硫均不利。当 $R = 3.5 \sim 4.2$ 时, 出钢脱硫效率 η_S 最高。但原始硫含量 $[S]_0$ 越低, 出钢脱硫效率 η_S 也越低。熔渣的碱度和流动性, 这两者之间的关系极为密切。碱度过低影响脱硫反应的顺利进行, 过高又破坏熔渣良好的流动性, 阻碍硫在渣中的扩散与转移。所以, 保持还原渣的合适碱度和良好的流动性是出钢过程强化脱硫操作的又一项措施。

(3) Aluminum for enhanced final deoxidation. When the amount of aluminum used for final

deoxidation is increased from 0.5~1kg/t steel, the dissolved oxygen in steel decreases, and the desulfurization efficiency of tapping steel also increases. Therefore, in order to better desulfurization, it is very important to strengthen the amount of aluminum used for final deoxidation, and the amount must meet the process requirements in the operation process. Aluminum added to steel can not be directly desulfurized, but can only assist in desulfurizing. Its main function is to reduce the oxygen content in steel. When the content of [%Al] ≥ 0.06% in steel, the oxygen content in steel will increase obviously, so the content of residual aluminum in general steel is 0.02% ~ 0.05%.

（3）强化终脱氧用铝。终脱氧用铝量由0.5kg/t钢增加到1kg/t钢时，钢中溶解氧降低，而出钢的脱硫效率也跟着提高，因此为了更好地脱硫，强化终脱氧用铝十分重要，在操作过程中用量一定要满足工艺要求。加入钢中的铝不能直接脱硫，只能协助脱硫，它的主要作用是为了降低钢中的氧含量。当钢中的［%Al］≥0.06%时，钢中的氧含量将明显增高，因此一般钢中均规定残余铝的含量为0.02%~0.05%。

(4) Strengthen the way of tapping. The desulfurization reaction is an interface reaction, and the expansion of the reaction interface is conducive to desulfurization. If the reaction interface between slag and steel is enlarged by the fierce mixing of slag and steel in the process of tapping, the desulfurization reaction conditions can be improved significantly. The general rule is that the desulfurization efficiency of tapping will increase with the increase of mixing speed, mixing height, slag amount, slag temperature and FeO content. In order to make full use of this opportunity, at present, EAF steelmakers usually adopt slag steel mixing, large orifice spraying and fast and powerful way of tapping, which can continue to desulfurate 30%~50%. Therefore, it is required that the tap hole be large and the tap hole be deep. For the overhead electric furnace, the design height of the platform is required to be higher.

（4）强化出钢方式。脱硫反应是界面反应，扩大反应界面有利于脱硫。如果在出钢过程中，利用渣钢的激烈搅混来扩大渣钢间的反应界面，可使脱硫反应条件得到显著的改善。一般规律是出钢脱硫效率将随混冲速度的提高、混冲高度的增加、渣量的加大、渣温的提高以及（FeO）含量的降低而提高。为了充分利用这样的机会，目前电炉炼钢工多采取渣钢混出、大口喷吐、快速有力的方式出钢，可继续脱硫30%~50%。为此，要求出钢口要大、出钢坑要深。对于高架式的电炉，要求平台的设计高度要高些。

3.4.3 Adjustment and Measurement of Liquid Steel Temperature
3.4.3 钢液温度的调整与测量

3.4.3.1 Effect of Temperature on Liquid Steel Refining
3.4.3.1 温度对钢液精炼的影响

The smelting temperature of EAF steel is mainly controlled in the oxidation period, while the reduction period is often adjusted as necessary. It is very important to control the smelting process by correctly controlling the temperature of reduction period, which can improve the quality and output of steel and reduce various consumption. If the smelting temperature is too high, the power consumption will increase, the reduction slag will become thin and yellow, which is difficult to control. At the same time, the deoxidization ability of the elements will be reduced, which will make the liquid steel inhale seriously or increase the total amount of inclusions in the steel. In addition, it also reduces the service life of furnace lining and increases the consumption of refractories. If the tapping temperature is too high, it is easy to break the plug rod, and once the plug rod is broken, it will inevitably affect the control of injection temperature and injection rate. High temperature pouring is easy to make steel running and die sticking, ingot (piece) is also easy to produce subcutaneous bubble, shrinkage cavity, hot crack, and it is easy to cause defects such as white spot or uneven carbide on the finished product. The viscosity of low-temperature smelting melt is large, the physical and chemical reaction between slag and steel can not be fully carried out, which not only prolongs the smelting time, but also affects the deoxidization, desulfurization effect and the floatation and removal of inclusions. The finished products are also prone to defects such as hairline or fracture. Low temperature pouring is easy to make the nozzle or the middle injection pipe coagulate and affect the normal process of pouring, and even cause waste products. Even the ingot which is just poured is easy to have defects such as peeling, scarring, cold cutting or loose center. The low pouring temperature and the poor quality of ingot surface not only increase the cleaning amount, but also waste human and material resources.

电炉钢冶炼温度的控制主要是在氧化期，还原期只是经常进行必要的调整。正确掌握还原期的温度能够很好地控制冶炼过程，这对提高钢的质量和产量及降低各种消耗等均有重要的作用。冶炼温度过高，电耗增加，还原渣易变稀变黄难以控制，同时也降低元素的脱氧能力，使钢液吸气严重或使钢中夹杂物总量增加。另外，也降低炉衬的使用寿命，增加耐火材料消耗。如果出钢温度过高，容易蚀断塞棒，而塞棒一旦蚀断就必然影响注温注速的控制。过高温度的浇注极易造成跑钢、粘模、锭（件）还易产生皮下气泡、缩孔、热裂，在成品材上容易引起白点或碳化物不均匀等缺陷。低温冶炼熔体的黏度大，渣钢间的物化反应不能充分地进行，既延长了冶炼时间，也影响脱氧、脱硫效果以及夹杂物的上浮与排除，成品材还容易出现发纹或断口等缺陷。低温浇注又易使水口或中注管凝死而影响浇注的正常进行，严重的甚至造成废品，就是勉强注成的钢锭也容易出现翻皮、结疤、冷截或中心疏松等缺陷。浇注温度低，钢锭表面质量差，既增加了清理量，也浪费了人力物力。

3.4.3.2 Adjustment of Liquid Steel Temperature
3.4.3.2 钢液温度的调整

The temperature system in the oxidation period creates conditions for smelting in the reduction period. The temperature adjustment in the reduction period is gradually completed on the basis of the temperature in the oxidation period, considering the characteristics of the steel, the tapping temperature, the characteristics of the refining outside the furnace and the pouring conditions.

氧化期的温度制度为还原期冶炼创造了条件，还原期温度的调整是在氧化期温度的基础上，考虑钢种的特点、出钢温度、炉外精炼的特点以及浇注条件等而逐渐完成的。

(1) Determination of tapping temperature. The tapping temperature is generally determined as follows:

$$\text{tapping temperature} = \text{casting temperature} + \text{tapping temperature drop} + \text{refining and sedation temperature drop}$$

（1）出钢温度的确定。出钢温度一般按下式确定：

$$\text{出钢温度} = \text{开浇温度} + \text{出钢温降} + \text{精炼与镇静温降}$$

Different steel grades have different characteristics, different performance and quality inspection standards, so the requirements of tapping temperature are also different. A large number of production practices have summarized that: high carbon steel has low melting point, good fluidity, low tapping temperature and high tapping temperature; for high viscosity steel, such as high chromium steel, the tapping temperature should be higher, while for high fluidity or high refractory corrosion, such as high silicon steel, high manganese steel or high manganese high silicon steel, the tapping temperature should be lower; the temperature control of high alloy steel with more alloying elements and impurities should be higher, while that of general alloy steel should be lower; the temperature control of steel with high sensitivity to defects such as hairline or fracture should be higher, while that of steel with high sensitivity to defects such as crack, shrinkage cavity, white spot or uneven carbides should be lower; the temperature of steel output at the time of injection should be lower than that at the time of injection; the tapping temperature of multi plate pouring should be higher than that of single plate pouring. For the refining outside the furnace or alloying in the ladle or adding solid synthetic slag in the ladle without heating capacity, the tapping temperature is higher; generally, the tapping temperature in summer is lower than that in winter; the tapping temperature of less tapping should be higher than that of more tapping.

不同的钢种具有不同的特点，也有不同的性能和质量检验标准，因而对出钢温度的要求也相应不同。大量的生产实践已总结出：高碳钢的熔点低、流动性好，出钢温度应低一些，低碳钢的出钢温度应高一些；黏度大的，如高铬钢等，出钢温度应高一些，而流动性较好或对耐火材料腐蚀作用大的，如高硅钢、高锰钢或高锰高硅钢，出钢温度应低一些；合金元素较多和杂质较多的高合金钢温度控制应高一些，而一般合金钢的温度控制相应要低一些；对发纹或断口等缺陷敏感性强的钢种，温度控制应高一些，而对裂纹、缩孔、白点或碳化物不均匀等缺陷敏感性大的钢种，温度控制应低一些；上注时的出钢温度应比下注时的低一些；多盘浇注的出钢温度应比单盘浇注的高一些，对于没有升温能力的炉外精炼或包中合金化或包中加固体合成渣的，出钢温度要求更高一些；一般夏季出钢温度要比冬季低一些；出钢量少的出钢温度应比出钢量多的高一些。

(2) Adjustment of liquid steel temperature. Since the molten steel has been heated to be equal to or slightly higher than the tapping temperature before reduction, and the melting point of the reduced steel is decreased, the adjustment process of the temperature in the reduction period is actually the process of keeping or gradually reducing the molten steel temperature to the tapping

temperature. The adjustment of temperature in reduction period mainly depends on the flexible and correct use of current and voltage, generally making smelting temperature gradually reduce from high to meet the requirements of tapping temperature. After full slagging, in order to make up for the loss of molten steel heat and the heat consumption of slagging materials and alloy melting during slagging, a large power should be supplied. After the formation of thin slags, the voltage or current should be reasonably reduced according to the specific situation, so as to avoid heating up or sudden temperature drop due to power failure after the reduction period. During the reduction period, the power cut and temperature drop are mostly caused by the high smelting temperature at the end of the oxidation period. The sudden power cut and temperature drop will make the molten pool lose the function of arc stirring and affect the uniformity of the composition and temperature of the molten steel. It will also make the slag suddenly thicken and deteriorate, weaken the reduction atmosphere in the furnace, and reduce various physical and chemical reactions between the slag and steel, such as deoxidation and desulfurization. At the same time, the rapid cooling will also cause great damage to the furnace lining. Therefore, when controlling the smelting temperature, the steelmakers should try their best to avoid the power cut and rapid cooling in the reduction period.

（2）钢液温度的调整。因钢液在还原前已被加热到等于或稍高于出钢温度，而还原后钢的熔点是下降的，所以还原期温度的调整过程实际上就是使钢液温度保持或逐渐下降到出钢温度的过程。还原期温度的调整主要是靠灵活正确地使用电流与电压，一般是使冶炼温度由高逐渐降低并达到满足出钢温度的要求为止。全扒渣后，为了弥补扒渣时钢液热量的散失与造渣材料及合金熔化的耗热，应供给较大的功率，稀薄渣形成后再根据具体情况，合理地降低电压或减少电流，以避免在还原期后升温或停电急剧降温。还原期的停电降温大多是氧化末期冶炼温度过高所致，停电急剧降温会使熔池失去电弧搅拌的作用而影响钢液成分和温度的均匀，还会使熔渣突然变稠变差而减弱炉中的还原气氛，降低渣钢间的各种物化反应如脱氧、脱硫等，同时因急冷也会给炉衬带来较大的损坏。所以，电炉炼钢工在控制冶炼温度时，应尽量避免还原期的停电急剧降温。

3.4.3.3 Measurement of Liquid Steel Temperature
3.4.3.3 钢液温度的测量

During the reduction period, the molten pool is relatively calm, and the temperature in each area is difficult to be uniform. The temperature of molten slag is the highest at the bottom of the electrode and the lowest near the slag line on the furnace wall; the temperature of molten steel is higher at the upper layer than that at the lower layer. When the ordinary power electric furnace is not stirred, the temperature difference between the upper layer and the lower layer can reach 50℃, which is ideal, and the difference is 10~20℃; the temperature difference between the molten steel near the slag line and the central part is about 30~40℃; the temperature of molten slag is generally higher than that of molten steel 40~80℃. Therefore, in order to make the temperature of liquid steel in reduction period tend to be uniform, necessary stirring should be carried out frequently, especially before temperature measurement.

还原期的熔池比较平静,各个区域的温度难以均匀。熔渣的温度在电极底部最高,靠炉壁渣线附近的较低;钢液温度,上层比下层的高,普通功率电炉在没有搅拌时,上下温差可达 50℃,搅拌得理想一些,还相差 10~20℃;渣线附近与中心部位钢液的温差为 30~40℃;熔渣温度一般高于钢液温度 40~80℃。因此,为使还原期钢液的温度趋于均匀,应经常进行必要的搅拌,尤其是在测温之前这项操作更显得突出重要。

(1) Instrument measurement of liquid steel temperature. There are the following methods to measure the temperature of molten steel:

(1) 钢液温度的仪表测量。钢液温度的仪表测量有以下方法:

1) Optical pyrometer. Optical pyrometer is also called colorimetric pyrometer. Its principle is to compare the resistance of pyrometer with the color of molten steel after current. When it is consistent, the temperature reading of pyrometer is the optical temperature of molten steel. This method is easy to measure, but the error is large. The measured result is not the actual temperature of molten steel. Generally, the optical temperature of molten steel is about 80~100℃ lower than that of thermal couple.

1) 光学高温计测温法。光学高温计又称为比色高温计。它的原理是将高温计的电阻丝通以电流后的颜色与钢液的颜色比较,当一致时,高温计的温度读数即是钢液的光学温度。此法测量简便,但误差大,测得结果也不是钢液的实际温度,一般钢液的光学温度比热电偶温度低 80~100℃。

2) Thermocouple thermometry. Thermocouples are made of different metals with different thermoelectric potentials at different temperatures. There are many kinds of thermocouples, however, they can be divided into point measurement and continuous measurement according to the different ways of use: point measurement usually uses expendable intrusive thermocouple, while continuous measurement is to place the continuous thermometer on a certain part of the furnace body or ladle for measurement. The value measured by thermocouple can basically reflect the actual temperature of molten steel, so it has been widely used. However, it should be pointed out that the results measured by the point measuring thermocouple are representative only when the operation is normal, the furnace condition is good and the furnace is fully stirred. When the furnace condition is bad, the MgO content in the slag is high or the temperature of the slag is raised later, sometimes the temperature of the slag is much higher than the temperature of the molten steel. At this time, the temperature of the molten steel measured by the thermocouple is often much lower than the temperature of the slag. However, through In the process of tapping, the heat of slag will be transferred to the molten steel and the temperature of molten steel will rise. At this time, the temperature measured in the furnace will be low, but the actual temperature of the molten steel in the ladle will increase after tapping.

2) 热电偶测温法。热电偶是利用不同的金属在不同的温度下具有不同的热电动势制成的。热电偶的种类较多,然而根据使用方式不同分为点测和连测两种:点测多使用消耗型侵入式热电偶,而连测是将连续测温仪安放在炉体或钢包的某一部位上进行测量。热电偶测得的数值基本上能反映钢液的实际温度,因此目前获得了比较广泛的应用。但有一点需指出,点测热电偶测得的结果是在操作正常、炉况良好和充分搅拌的情况下才具有代表性,而当炉况差、熔渣中 MgO 的含量较高或在后升温的情况下,熔渣的温度有时远远高于钢液的温度,这时热电偶所测的钢液温度也就往往低于熔渣温度很多,可是通过出钢过程的渣钢混冲,熔渣的热量就要传给钢液而使钢液的温度升高。这时候就会出现炉内测量的温度低,而出钢后钢包内钢液的实际温度却增高的现象。

(2) Experience judgment of liquid steel temperature. The temperature of the liquid steel is mainly measured by the instrument:

(2) 钢液温度的经验判断。钢液的温度主要靠仪表来测量,除此之外,电炉炼钢工还经常利用下述经验进行判断:

1) Judgment method of liquid steel conjunctiva (static membrane). Different kinds of steel have different surface conjunctiva (static film) temperature. The higher the temperature of the liquid steel is, the longer the time required for it to fall to the temperature of the conjunctiva (static film). Therefore, the temperature of the liquid steel can be indirectly judged according to the time of the conjunctiva (static film) of the liquid steel. The conjunctival (static) time of molten steel is generally measured in seconds. Carbon structural steel and carbon tool steel shall be subject to the conjunctive, and high aluminum or high chromium steel shall be subject to the static film. See Table 3-4 for the reference relationship between the number of seconds of the conjunctiva (static membrane) of the liquid steel and the corresponding approximate temperature.

1) 钢液结膜(静膜)判断法。不同钢种的钢液具有不同的表面结膜(静膜)温度。钢液的温度越高,下降到结膜(静膜)温度所需的时间越长,因此根据钢液的结膜(静膜)时间可以间接地判断钢液的温度。钢液的结膜(静膜)时间一般用秒计量。碳素结构钢和碳素工具钢以结膜为准,高铝或高铬钢等以静膜为准。钢液的结膜(静膜)秒数与相对应的大概温度参照关系见表3-4。

Table 3-4 Reference relationship between seconds of liquid steel conjunctive (static film) and liquid steel temperature

表 3-4 钢液结膜(静膜)秒数与钢液温度的参照关系

Conjunctival time/s 结膜时间/s	Optical pyrometer temperature/℃ 光学高温计温度/℃	Thermocouple (Mo-W) temperature/℃ 热电偶(Mo-W)温度/℃
15~22	1435~1440	1500~1520
22~27	1440~1450	1520~1545
27~30	1450~1460	1545~1555
30~32	1460~1465	1555~1565
32~33	1465~1470	1565~1575
33~35	1470~1480	1575~1585
35~36	1480~1485	1585~1590

Continued Table 3-4

Conjunctival time/s 结膜时间/s	Optical pyrometer temperature/℃ 光学高温计温度/℃	Thermocouple (Mo-W) temperature/℃ 热电偶 (Mo-W) 温度/℃
36~37	1485~1490	1590~1595
37~38	1490~1500	1595~1610
38~39	1500~1510	1610~1620
39~40	1510~1520	1620~1630
40~42	1520~1530	1630~1640
>42	>1530	>1650

Note: this table is applicable to carbon structural steel, alloy structural steel, carbon tool steel, rounding tool steel and spring steel.

注：此表适用于碳素结构钢、合金结构钢、碳素工具钢、合金工具钢、弹簧钢。

It is a simple and easy method to judge the temperature of molten steel by using the time of conjunctive (static film) of molten steel, which has been widely used in production. However, this method is often affected by production or external conditions, and sometimes it is difficult to accurately reflect the real temperature of the molten steel, such as under the effect of cold air flow, or the slag condition is poor, or the slag color is abnormal, or the operation is not standard (such as no slag cover or less molten steel), the heat dissipation of the molten steel in the scoop is fast, and the number of seconds of the film (static film) does not look high, while the actual temperature may be high, and some sample scoops are repeatedly. When using this method to judge, do not be confused by these illusions. Because the volume of scoop and the thickness of scoop wall have a direct impact on the time of conjunctiva (static membrane) of molten steel, it is very important to choose a standard scoop. At present, the size of the commonly used sample scoop is shown in Figure 3-3, and the weight of the scoop when it is full of liquid steel is about 1kg.

利用钢液的结膜（静膜）时间来判断钢液温度的高低是一种简单易行的方法，在生产上得到了广泛的应用。但该法往往受生产或外界条件的影响，有时也难以确切反映钢液的真实温度，如在冷空气流的作用下，或渣况不良、渣色不正常，或操作不标准（如没有熔渣覆盖或钢水量少），钢液在勺内散热快，结膜（静膜）秒数看上去不高，而实际温度可能高，也有的将样勺反复多次粘渣或用红勺盛取高温钢液，或样勺容量尺寸超标准，结膜秒数很高，而实际温度可能不高，因此利用该法进行判断时，不要被这些假象所迷惑。由于样勺的容积大小、勺壁的厚度对钢液的结膜（静膜）时间有直接影响，因此选用标准的样勺是很重要的。目前，一般常用样勺的尺寸如图3-3所示，该种勺盛满钢液时的重量约为1kg。

2) Observe the judging method of liquid steel color. Because the liquid steel has different colors at different temperatures, we can judge the temperature according to the color of the liquid

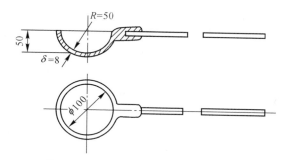

Figure 3-3　dimensions of sample scoop
图 3-3　样勺的尺寸

steel. This method is more representative after proficiency. When observing the liquid steel in the scoop:

①The molten steel is red or dark red, and the temperature is very low, about 1550℃;
②The molten steel is bright red and the temperature is about 1600℃;
③The molten steel is blue and white, and the liquid surface has smoke seedlings, about 1620℃;
④The molten steel is white and smoke, about 1650℃;
⑤The molten steel turns white completely, feels dazzling under the blue glasses, and the thick smoke emits directly, indicating that the temperature is very high, about 1670℃ above.

2) 观察钢液颜色判断法。由于钢液在不同的温度下具有不同的颜色,可以根据钢液的颜色来判断温度的高低。该法熟练后比较有代表性。在样勺中观察钢液时:

①钢液呈红色或暗红色,温度很低,约为1550℃以下;
②钢液呈亮红色,温度约为1600℃;
③钢液呈青白色,液面带有烟苗,约为1620℃;
④钢液呈白色,冒浓烟,约在1650℃;
⑤钢液完全发白,在蓝眼镜下感到耀眼,且浓烟直冒,表明温度很高,约在1670℃以上。

3) Temperature measurement with ladle. In this method, several sample spoons are used to take the liquid steel, and the liquid steel is poured out after standing for different seconds. After observing how many seconds the liquid steel starts to stick the spoon, and the number of seconds when the spoon starts to stick is taken as the sign of the temperature of the liquid steel, which is called the number of seconds to stick the spoon. This method is often used in alloy steel with high chromium, aluminum and titanium.

3) 钢液粘勺测温法。该法是用几个样勺盛取钢液,分别静置不同的秒数后将钢液倒出,观察钢液经过多少秒后开始粘勺,并以开始发生粘勺的秒数作为钢液温度的标志,称为粘勺秒数。此法经常用于含高铬、高铝、高钛的合金钢上。

4) Determination method of sample solidification state. Generally, the molten steel is slowly injected into the sample bowl. If the shrinkage is severe and the edge is thin, it means the temperature is high; if it is round, the temperature is medium; if it is convex, it means the

temperature is low.

4）试样凝固状态判断法。一般是将钢液慢慢地注入试样碗内，如收缩很厉害，边缘呈尖薄，表示温度高；呈圆形，温度中等；凸起，则表明温度很低。

5) Judgment method of molten steel flow. Slowly pour the molten steel on the smooth and clean iron plate from a height of about 10cm. The longer the molten steel flows, the higher the temperature is; if the iron plate is welded, the higher the temperature is. This method is often used in high alloy steel with low melting point or more oxidizable elements, which cannot be judged by other experience.

5）钢液流动情况判断法。将钢液从10cm左右的高度上慢慢地浇在光滑、清洁的铁板上，钢液流动的距离越长，说明温度越高；如果将铁板焊住则表明温度更高。该法常用于熔点低或易氧化元素较多而用其他经验无法判断的高合金钢上。

6) The method of judging the corrosion of steel bars. Bend the 10~12mm steel bar into obtuse angle, insert it into the ladle and stir it back and forth, and withdraw it for inspection after 5~8s. If the section of the steel bar is thin and sharp, it means that the temperature is high; if it is flat and rough, it means that the temperature is general; if there is residual steel on the surface of the steel bar, it means that the temperature is low.

6）钢条熔蚀判断法。将10~12mm的钢条弯成钝角，插入样勺的钢液中来回搅动，5~8s后抽出检查。如果钢条断面细而尖，说明温度较高；平而粗糙表示温度一般；钢条表面粘有残钢则表明温度较低。

In addition, the temperature of the liquid steel can be roughly judged from the composition of some elements: under normal operation, manganese is the symbol of the temperature of the liquid steel, if there is no loss of manganese content in the steel at the end of oxidation, it indicates that the actual temperature of the liquid steel is higher; if the amount of sulfur removal of the steel at the end of reduction is larger or the amount of silicon powder is normal, but the yield rate is low, it also indicates that the actual temperature of the liquid steel is higher. It can also be judged from the reduction atmosphere: if the electrode hole is not closed before tapping, or if there is a sudden fire when tapping, sometimes there is a 'ga la, ga la' sound in the molten pool, indicating that the actual temperature of the molten steel is not high; if the electrode hole has been closed automatically, but there is a rolling smoke from the corner seam of the furnace door, or the flame is large when adding silica powder, indicating that the temperature is high. Of course, it can also be judged from whether there is boiling near the slag line: if the slag is thin at the end of reduction or there is boiling near the slag line, it means that the temperature of the molten steel is high, otherwise it is only general.

此外，还可从某些元素的成分上对钢液温度进行大概的判断：在操作正常的情况下，锰是钢液温度的标志，如果氧化末期钢中锰含量没有损失，表明钢液的实际温度较高；如果还原末期钢的脱硫量较大或硅粉用量正常，但收得率低，也说明钢液的实际温度较高。还可从还原气氛上进行判断：如果出钢前电极孔没有封闭死，或时出时缩突突地冒出火苗，有时熔池中还发出"嘎啦、嘎啦"的响声，表明钢液的实际温度不高；如果电极孔已自动封闭死，但却从炉门的边角缝处冒出滚滚的浓烟，或加硅粉时火焰大，说明温度较高。当然，还可从渣线附近是否沸腾进行判断：如果还原末期熔渣稀或渣线附近处有沸腾，说明钢液温度较高，否则只是一般。

3.4.4 Adjustment of Liquid Steel Composition
3.4.4 钢液成分的调整

3.4.4.1 Addition Principle and Yield of Alloy Elements
3.4.4.1 合金元素的加入原则及收得率

Addition principle of alloy elements:
(1) The alloy elements should be melted quickly and evenly distributed in the molten steel;
(2) High yield and low cost;
(3) Impurities, such as SiO_2 and Al_2O_3, brought into molten steel by alloy materials can be removed;
(4) The added alloy material should not fluctuate too much to the bath temperature, even if the fluctuation is large, it should also be mastered by the steelmaker, otherwise it will affect the normal operation.

合金元素的加入原则：
(1) 要使合金元素在钢液中快速熔化、分布均匀；
(2) 收得率要高，成本要低；
(3) 合金材料带入钢液中的杂质，如 SiO_2、Al_2O_3 等能有机会去除；
(4) 加入的合金材料对熔池温度不要波动过大，即使波动较大，也应在炼钢工的掌握之中，否则将会影响操作的正常进行。

In the smelting process, in order to meet the above-mentioned adding principle, the adding time and method should be determined according to the physical and chemical properties, use amount and smelting method of alloy elements. If the affinity between alloy elements and oxygen is smaller than that of iron, such as nickel, etc., it should be added in the charging or melting period when it is used in large quantity, and in the oxidation or reduction period when it is used in small quantity. Alloy elements with the same affinity as oxygen and iron ratio, such as tungsten, shall be added in the charging or melting period when smelting by back blowing method or non oxidation method, such as smelting by ore oxygen method, at the end of oxidation or early reduction; alloy elements with a slightly higher affinity than iron, such as Cr, Mn, etc., shall be added after removing all oxide slag or reduction period; alloy elements with oxygen easily oxidized elements with higher affinity than iron, such as Ti, B, etc., should be added before or during tapping. The refractory alloy elements with high melting point should be transferred in as early as possible under the condition of considering their oxidation properties. Although some alloy elements have higher affinity with oxygen than iron, they can also be loaded with furnace charge, such as ferrochrome, etc. In order to facilitate the operation in front of the furnace, the adding time, adding method and yield of some alloy materials are detailed as follows:

在冶炼过程中，为了满足上述的加入原则，应根据合金元素的物化性质、使用量和冶炼方法等来确定加入时机和加入方法。合金元素与氧的亲和力比铁小的，如 Ni 等，大量使用时，应在装料或熔化期加入，少量使用时，应在氧化期或还原期调入。与氧的亲和力和铁比不相上下的合金元素，如 W 等，如采用返吹法或不氧化法冶炼时，应在装料或熔化期加入，如采用矿氧法冶炼时，应在氧化末期或还原初期加入；与氧的亲和力比铁稍大一些的合金元素，如 Cr、Mn 等，应在全部扒除氧化渣后或还原期加入；与氧的亲和力比铁更大的易氧化元素，如 Ti、B 等，应在出钢前或出钢过程中加入。对于高熔点的难熔合金元素，在考虑其氧化性能的条件下，应尽早一些调入。对于采用返吹法或不氧化法冶炼的高合金钢，尽管有的合金元素与氧的亲和力比铁大，但它们也可随炉料一同装入，如铬铁等。为便于炉前操作，现将一些合金材料的加入时间、加入方法及其收得率详述如下：

(1) Nickel. Under steel-making conditions, nickel does not actually oxidize. When used in large quantities, it can be loaded with the furnace charge, or in the melting period or oxidation period. In this way, the gas brought in by nickel can be well removed after oxidation boiling, while for the nickel used in small quantities, it is generally transferred in the reduction period. Due to the volatilization loss of nickel under the high temperature of the arc, it should be installed far away from the high temperature area of the arc when it is loaded with the furnace charge. In general, when smelting low nickel alloy steel, the yield of nickel can be calculated as 100%, while for smelting nickel base alloy, the evaporation loss can reach 2% ~ 3%.

(1) 镍。在炼钢条件下，镍实际上不氧化。大量使用时可随炉料一同装入，也可在熔化期或氧化期加入，这样经过氧化沸腾能使镍带入的气体得以较好的排除，而对于少量使用的镍一般在还原期调入。由于镍在电弧高温下有挥发损失，因此当随同炉料装入时，应装在远离电弧高温区的部位。一般冶炼低镍合金钢时，镍的收得率可按 100% 计算；而对于冶炼的镍基合金，挥发损失可达 2% ~ 3%。

(2) Cobalt and copper. Under the condition of steelmaking, cobalt and copper are also non oxidizing elements. They can be loaded with the furnace charge or adjusted with the reduction period. The recovery rate is considered as 100%.

(2) 钴和铜。在炼钢条件下，钴和铜也是属于不氧化元素，既可随炉料一同装入，也可在还原期随用随调，收得率均按 100% 考虑。

(3) Ferromolybdenum. Ferromolybdenum has high melting point, high density, and MoO_3 is volatile. Ferromolybdenum is usually loaded with the charge, and can also be added at the end of melting or oxidation or under the thin slag. The ferromolybdenum added at the end of reduction should be small and added 15 minutes before tapping. The yield of ferromolybdenum is generally 95% ~ 100%, but when the content of molybdenum in steel is more than 4%, the yield should be taken as the lower limit.

(3) 钼铁。钼铁的熔点高，密度大，MoO_3 易挥发。钼铁一般随炉料一同装入，也可在熔化末期、氧化末期或稀薄渣下加入。还原末期补调的钼铁应选用小块的并在出钢前 15min 加入。钼铁的收得率一般为 95% ~ 100%，但当钢中钼含量大于 4% 以上时，收得率应取下限。

(4) Calcium molybdate. When smelting molybdenum steel with molybdenum content less than 1%, calcium molybdate ($CaMoO_4$) can be used instead of ferromolybdenum. It is either loaded with the charge or added during the oxidation period. During smelting, calcium molybdate can react with iron and carbon as follows:

$$CaMoO_4 + 3[Fe] = CaO + 3(FeO) + [Mo]$$
$$CaMoO_4 + 3[C] = CaO + 3\{CO\} + [Mo]$$

（4）钼酸钙。冶炼钼含量小于1%的钼钢时，可用钼酸钙（$CaMoO_4$）代替钼铁，它或随炉料装入或在氧化期加入。在冶炼过程中，钼酸钙与铁和碳能发生如下反应：

$$CaMoO_4 + 3[Fe] = CaO + 3(FeO) + [Mo]$$
$$CaMoO_4 + 3[C] = CaO + 3\{CO\} + [Mo]$$

Because of the complete reduction of molybdenum in calcium molybdate, there is no oxide of molybdenum in the slag. However, in order to ensure the reduction performance of the molten steel and the stability of molybdenum content, calcium molybdate is better not to be used as the supplementary composition before tapping.

由于钼酸钙中的钼还原很完全，因此熔渣中不存在钼的氧化物。但为了保证钢液的还原性能和钼含量的稳定，钼酸钙最好不要用于出钢前的补调成分。

(5) Ferrotungsten. Compared with nickel and molybdenum, tungsten has higher affinity with oxygen. When the content of (FeO) is higher, it has oxidation loss. In addition, tungsten oxide volatilizes at high temperature. In order to reduce the loss of tungsten, ferrotungsten should be added to the steel at the end of oxidation or under the thin slag. For tungsten steel smelted by back blowing method or non oxidation method, the ferrotungsten can be loaded together with the furnace charge, and then adjusted at the end of oxidation or under the thin slag. The ferrotungsten supplemented at the end of reduction shall be added 15 minutes before tapping. Due to the high density and melting point of ferrotungsten, it is easy to deposit and melt slowly at the bottom of the furnace. Therefore, preheating is required before adding, and full stirring is required after adding and before tapping to ensure that the tungsten content in the molten steel tends to be uniform. The yield of tungsten is about 85% ~ 98%, and the higher the content is, the higher the yield is.

（5）钨铁。钨和镍、钼相比，与氧的亲和力较大，当（FeO）的含量较高时，它有氧化损失。此外，钨的氧化物在高温下还要挥发。为了减少钨的损失，氧化法冶炼的钨钢，钨铁应在氧化末期或稀薄渣下加入。采用返吹法或不氧化法冶炼的钨钢，钨铁可随炉料一同装入，然后在氧化末期或稀薄渣下调整，还原末期补调的钨铁应在出钢前15min加入。由于钨铁密度大，熔点高，易沉积炉底熔化慢，因此加入前要预热，加入后和出钢前要充分搅拌，以保证钢液中钨含量趋于均匀。钨的收得率一般为85%~98%，且含量越高，收得率也越高。

(6) Ferroniobium. The affinity between niobium and oxygen is higher than that of iron. When smelting by oxidation method or back blowing method, ferroniobium is added 20 ~ 30 minutes before tapping. If the smelting temperature of the process is low, it needs more than 40 minutes to make the melting complete. When smelting by non oxidation method, ferroniobium can be charged together with the charge. When deoxidization is good, the recovery rate of niobium is generally

95%~100%。

(6) 铌铁。铌与氧的亲和力大于铁。氧化法或返吹法冶炼时，铌铁在出钢前20~30min内加入，如果工艺的冶炼温度较低，则需40min以上，以使熔化完全。不氧化法冶炼时，铌铁可随炉料一同装炉，脱氧良好时，铌的收得率一般为95%~100%。

(7) Ferrochrome. The affinity between chromium and oxygen is higher than that of iron. The ferrochrome used in oxidation smelting is added under the thin slag; the ferrochrome produced by back blowing or non oxidation smelting can be loaded together with the furnace charge; the ferrochrome supplemented at the end of reduction is added 10 minutes before tapping. If the capacity of the furnace is small and a large amount of ferrochrome is added before tapping, the smelting time should be prolonged accordingly, so as to facilitate the melting of ferrochrome and the temperature rise of the molten pool. The yield of ferrochromium in general steel is about 95%~98%, while that of high chromium steel is about 80%~90%.

(7) 铬铁。铬与氧的亲和力大于铁。氧化法冶炼所用的铬铁在稀薄渣下加入；返吹法或不氧化冶炼的铬铁可随炉料一同装入；还原末期补调的铬铁在出钢前10min加入。容量小而在出钢前补加大量铬铁的炉子，要相应延长冶炼时间，以利于铬铁的熔化与熔池的升温。一般钢中铬铁的收得率为95%~98%，而用返吹法冶炼高铬钢时，收得率为80%~90%。

(8) Ferromanganese. The affinity between manganese and oxygen is higher than that of iron. Ferromanganese shall be added after the formation of thin slag or with slag material, and the adjusted ferromanganese at the end of reduction shall be added 10 minutes before tapping. The yield of ferromanganese is generally 95%~100%.

(8) 锰铁。锰与氧的亲和力大于铁。锰铁应在稀薄渣形成后或随同渣料加入，还原末期调整的锰铁在出钢前10min加入。锰铁的收得率一般为95%~100%。

(9) Ferrovanadium. Vanadium is easy to oxidize and reduce, and the adding time of FEV should be determined according to the steel grade. When smelting low vanadium steel $[w(V) \leq 0.30\%]$, it should be added 8~15 minutes before tapping; when smelting medium vanadium steel $[w(V) = 0.30\% \sim 0.50\%]$, it should be added 20 minutes before tapping; when smelting high vanadium steel $[w(V) = 0.50\% \sim 1.00\%]$, it should be added 30 minutes before tapping; when smelting higher vanadium steel $[w(V) > 1\%]$, it can be added 40 minutes before tapping and adjusted before tapping. Ferrovanadium used in single slag smelting can also be added in batches at the end of melting. The yield of FEV is generally considered as 95%~98%.

(9) 钒铁。钒易于氧化也易于还原，钒铁的加入时间应根据钢种而定。冶炼低钒钢 $[w(V) \leq 0.30\%]$ 时，应在出钢前8~15min加入；冶炼中钒钢 $[w(V) = 0.30\% \sim 0.50\%]$ 时，应在出钢前20min加入；冶炼高钒钢 $[w(V) = 0.50\% \sim 1.00\%]$ 时，应在出钢前30min加入；炼制更高的钒钢 $[w(V) > 1\%]$ 时，钒铁可在出钢前40min加入并在出钢前进行补调。单渣法冶炼所用的钒铁也可于熔化末期分批加入。钒铁的收得率一般按95%~98%考虑。

(10) Ferrosilicon. The affinity of silicon and oxygen is large. Ferrosilicon is usually added under the condition of good deoxidation except for the mirror surface. A small amount of ferrosilicon

shall be added 10 minutes before tapping, and the yield shall be calculated as 100%. When smelting silicon steel, ferrosilicon should be added 10~20 minutes before tapping, and at the same time, a proper amount of lime should be added. At this time, the yield of silicon is about 95% ~ 98%.

（10）硅铁。硅和氧的亲和力较大。硅铁除镜面加入的外，一般均在脱氧良好的情况下加入。对于少量调入的硅铁应在出钢前10min加入，收得率按100%计算。冶炼硅钢时，硅铁应在出钢前10~20min加入，与此同时还要补加适量的石灰，这时硅的收得率为95%~98%。

（11）Ferrotitanium. Titanium is easy to oxidize, but the price of ferrotitanium is also relatively expensive, so it is mainly used for the alloying of steel, and when the liquid steel deoxidizes well, it is added 5~15 minutes before tapping. The density of ferrotitanium is small. After adding ferrotitanium, it is necessary to press it into the molten steel with an iron rake to increase the recovery of titanium from the molten steel. When a large amount of ferrotitanium is added, for example, the slag is thin, the temperature is high or the silicon in steel is also high, a part of the slag should be removed before adding, which can not only improve the yield of titanium, but also prevent silicon from exceeding the specification. Of course, slagging is not necessary, but the phenomenon of silicon increase in molten steel after adding ferrotitanium can not be ignored, because ferrotitanium contains higher aluminum and silicon, while Al and Ti will react with (SiO_2) in slags to reduce silicon. The yield of ferrotitanium added into the furnace is related to the deoxidation of molten steel and the titanium content in the steel. If the deoxidation is good, that is to say, FeO in slag is less than 0.5%, when the titanium content in steel is less than or equal to 0.15%, the yield is about 30%~50%; when the titanium content in steel is 0.20%~0.80%, the yield is about 50%~65%; when the titanium content in steel is more than 0.80%, the yield is 70%~90%.

（11）钛铁。钛易氧化，但钛铁的价格也比较贵，所以它主要用于钢的合金化，并在钢液脱氧良好的情况下，在出钢前5~15min加入。钛铁的密度较小，加入后要用铁耙子压入钢液中，以增加钢液对钛的收得率。当大量加入钛铁时，如熔渣稀，温度高或钢中硅也高，应先扒出一部分熔渣后再加入，这样既可提高钛的收得率，又可防止硅超出规格。当然也可不扒渣，但钛铁加入后钢液增硅的现象不容忽视，这是因为钛铁中含有较高的铝和硅，而Al、Ti还会与渣中的（SiO_2）发生反应使硅还原。炉中加入的钛铁的收得率与钢液的脱氧情况及钢中的钛含量有关。如果脱氧良好，即渣中$w(FeO)<0.5\%$，当钢中的钛含量小于或等于0.15%时，收得率为30%~50%；当钢中的钛含量为0.20%~0.80%时，收得率为50%~65%；当钢中的钛含量大于0.80%时，收得率为70%~90%。

（12）Aluminum ingot. The alloy of aluminum is made of aluminum ingot. In general, under the condition of good deoxidization, a proper amount of reducing slag should be removed 8~15 minutes before tapping and added after making new slag. In order to improve the yield of aluminum ingot, the iron rake should be used to beat the ingot continuously after adding to make the ingot sink and fully contact with the molten steel. The yield of aluminum ingot added into the furnace is related to the deoxidation of molten steel and the aluminum content in the steel. If deoxidation is good, that is to say, FeO in slag is less than 0.50%: when the aluminum content is 1%, the yield is 70% ~ 80%; when the aluminum content is 2%, the yield is 80% ~ 85%; when the aluminum content is 3%, the yield is 85% ~ 90%; when the aluminum content is more than 5%, the yield is 90% ~ 95%. Before the aluminum ingot is added, it is necessary to control the temperature of the molten steel, not only to create conditions for the floatation of the deoxidizing product Al_2O_3 of aluminum, but also to prevent the spot segregation, hairline, slate fracture and other defects of aluminum steel, as well as the high-temperature steel that cannot be poured.

（12）铝锭。铝的合金化多用铝锭。一般在脱氧良好的情况下在出钢前 8~15min 扒出适量的还原渣并补造新渣后加入。为了提高铝锭的收得率，加入后应用铁耙子不断地拍打，使铝锭下沉并与钢液充分接触。炉中加入的铝锭的收得率与钢液的脱氧情况及钢中的铝含量有关。如脱氧良好，即渣中的 $w(FeO)<0.50\%$：当铝含量（质量分数）为 1% 时，收得率为 70%~80%；当铝含量为 2% 时，收得率为 80%~85%；当铝含量（质量分数）为 3% 时，收得率为 85%~90%；当铝含量（质量分数）大于 5% 时，收得率为 90%~95%。铝锭加入前，要注意控制钢液的温度，既要为铝的脱氧产物 Al_2O_3 的上浮创造条件，进而有利于防止铝钢产生点状偏析和发纹及石板状断口等缺陷，同时也要避免出现无法浇注的高温钢。

（13）Ferroboron. The affinity of boron with oxygen and nitrogen is very strong. Before the addition of boron and iron, the steel must be deoxidized and denitrified well. Before addition, aluminum 1kg/t steel and titanium 0.05% (without burning loss) must be inserted to fix the oxygen and nitrogen in the steel. Ferroboron can be used in both furnace and ladle. When it is added into the furnace, the ferroboron shall be wrapped with aluminum sheet and inserted into the molten steel, and the steel shall be tapped within 3~5 minutes after it is added. When it is used in the ladle, slag shall be blocked and steel shall be mixed after adding. In order to reduce the loss of boron, some steels are added with 0.5~1kg/t silicon calcium before tapping. The yield of ferroboron is 30%~50%. If the wire feeder is used to feed boron into the molten steel, the yield can reach 90%.

（13）硼铁。硼与氧和氮的亲和力很强，硼铁加入前，钢液必须脱氧、脱氮良好，并在加前还要插铝 1kg/t 钢和加钛 0.05%（不计烧损），以固定钢液中的氧和氮。硼铁既可在炉中加入，也可在包中使用。在炉中加入时，硼铁要用铝皮包住并插入钢液中，加入后 3~5min 内出钢。包中使用时，应先挡渣出钢，加完后方可渣钢混出。为了减少硼的损失，有的在出钢前加硅钙 0.5~1kg/t 钢。硼铁的收得率为 30%~50%，如果采用喂丝机向钢液中喂硼，收得率可达 90%。

（14）Nitrogen. Nitrogen is easy to diffuse and escape, and cannot be added in the oxidation period. In the process of back blowing smelting, the yield of nitrogen is also low when the nitrogen containing return material is used, generally about 30%. When nitrogen is blown into the molten steel, although nitrogen content can also be increased, the yield is low and unstable. In electric furnace steelmaking, nitrogen, which is used as alloying element, is usually added in reduction period in the form of nitrogen manganese alloy or nitrogen chromium alloy. Generally, when the nitrogen content in the alloy is low (such as about 1%), the yield can reach 100%; when the nitrogen content is high (such as 6% ~ 7%), the yield is low. When the temperature of molten steel is too high, part of nitrogen in the nitrogen alloy is easy to volatilize, while when the temperature of molten steel is too low, the solubility of nitrogen decreases, both of which affect the yield. When P_{N_2} = 0.1MPa, Mn, Cr and Mo elements in the steel will not form independent phase nitride at the existing concentration in the steel, but can significantly improve the solubility of nitrogen in the liquid steel and greatly improve the yield. Therefore, in order to improve the yield of nitrogen, sufficient Mn, Cr or Mo should be allowed in the liquid steel before the addition of nitrogen alloy within the allowable range of steel specification. Oxygen is a strong surface active material in molten steel. When the oxygen content in steel is high, it also affects the dissolution rate of nitrogen. Therefore, nitrogen alloy should be added in the case of good deoxidation. In addition, the yield of nitrogen is also related to the smelting method. Generally, the yield of non oxidation smelting is higher and more stable than that of oxidation or back blowing smelting.

（14）氮。氮易于扩散逸出，不能在氧化期加入。在返吹法冶炼中，搭用含氮返回料时氮的收得率也较低，一般为30%左右。当向钢液中吹入氮气时，虽然也能增加氮含量，但收得率较低且不稳定。在电炉炼钢上，作为合金化元素使用的氮，通常以氮锰合金或氮铬合金的形式在还原期加入，影响收得率的因素主要有合金中的氮含量、钢液的温度及钢中的化学成分等。一般是合金中的氮含量低（如1%左右）时，收得率可达100%；氮含量（质量分数）较高（如6%~7%）时，收得率较低。钢液的温度过高时，氮合金中的部分氮易挥发；而钢液的温度过低时，氮的溶解度下降，两者均影响收得率。当P_{N_2} = 0.1MPa时，钢中的Mn、Cr、Mo元素在钢中现有的浓度下不会生成独立相的氮化物，但能显著地提高氮在钢液中的溶解度，而使收得率大为提高。因此，为了提高氮的收得率，在氮合金加入前，在钢种规格允许的范围内，尽量让钢液中先有足够的Mn、Cr或Mo。氧是钢液的强表面活性物质，当钢中的氧含量高时，也影响氮的溶解速率，所以氮合金尽量在脱氧良好的情况下加入。此外，氮的收得率还与冶炼方法有关，通常是不氧化法冶炼的收得率比氧化法或返吹法的高且又稳定。

(15) Sulfur and phosphorus. For smelting high sulfur steel, sulfur is added in the form of sulfur or sheet ferrous sulfide. Sulfur is added after fully raking slag, and the recovery rate is 50% ~ 70%; if it is used for powder spraying in ladle, the recovery rate is higher; sheet ferrous sulfide is added before tapping, and the recovery rate is 100%. In the smelting of phosphorus steel, phosphorus is added in the form of ferrophosphorus in the early or late reduction stage, and the yield is 100%. In order to ensure the recovery of sulfur and phosphorus, neutral slag system should be used in the reduction period.

（15）硫和磷。冶炼高硫钢，硫是以硫黄或片状硫化亚铁的形式加入。硫黄在全扒渣后加入，收得率为50%~70%；如用于包中喷粉，收得率更高；片状硫化亚铁在出钢前加入，收得率为100%。冶炼磷钢，磷是以磷铁的形式在还原初期或还原末期加入，收得率为100%。为了保证硫、磷的回收，还原期应采用中性渣系冶炼。

(16) Rare earth metals. Rare earth metals should be added into the furnace after final deoxidation, and the shorter the time of staying in the furnace, the higher the yield, generally about

20%~40%. Of course, the addition of rare earth metals into the ladle can also obtain satisfactory results.

(16) 稀土金属。稀土金属应在终脱氧后加入炉中效果较好,而且在炉内停留的时间越短,收得率越高,一般为20%~40%,当然,包中加入也能获得较为满意的结果。

(17) Rare earth oxides. Rare earth oxides are usually mixed with sodium nitrate or calcium boron or calcium silicon and then added into the ladle. For example, calcium boron is only used when boron is allowed in steel, and the yield is 30%~40%. Rare earth oxide has a strong ability to absorb hydrogen. Do not put it in the air for too long before use, so as not to bring the gas into the steel.

(17) 稀土氧化物。稀土氧化物一般是与硝酸钠或硼钙或硅钙混合后加入包中,如钢中允许硼存在时才使用硼钙,收得率为30%~40%。稀土氧化物吸收氢的能力较强,用前切勿在空气中放置过久,以免将气体带入钢中。

The addition time and yield of some commonly used alloy materials are shown in Table 3-5.
一些常用合金材料的加入时间和收得率见表3-5。

Table 3-5 Addition time and yield of common alloy materials
表3-5 常用合金材料的加入时间和收得率

Alloy name 合金名称	Smelting method 冶炼方法	Joining time 加入时间	Yield rate/% 收得率/%
Nickel 镍		charge 装料	>95
		Oxidation period and reduction period 氧化期、还原期	95~100
Molybdenum iron 钼铁		Charging, melting, oxidation, reduction period adjustment 装料、熔化末、氧化末、还原期调整	95~100
Tungsten iron 钨铁	Oxidation method 氧化法	End of oxidation, beginning of reduction 氧化末、还原初	85~95
	Blowback method 返吹法	Loading, reduction period adjustment 装料、还原期调整	85~90
	Non oxidation method 不氧化法	Same back blowing method 同返吹法	92~98
Niobium iron 铌铁	Various methods 各种方法	When deoxidation is good, add 20~40min before tapping 脱氧良好时,出钢前20~40min加入	95~100
	Non oxidation method 不氧化法	Loading, reduction period adjustment 装料、还原期调整	
Chromium iron 铬铁	Oxidation method 氧化法	Primary reduction 还原初	95~98
	Blowback method 返吹法	Loading, reduction period adjustment 装料、还原期调整	80~90
	Non oxidation method 不氧化法	Same back blowing method 同返吹法	80~90

Continued Table 3-5
续表 3-5

Alloy name 合金名称	Smelting method 冶炼方法	Joining time 加入时间	Yield rate/% 收得率/%
Ferromanganese 锰铁		Primary reduction 还原初 Before tapping 出钢前	95~98 约 100
Ferrovanadium 钒铁	Single slag method 单渣法	$w(V) \leqslant 0.30\%$, 8~15min before tapping $w(V) \leqslant 0.30\%$, 出钢前 8~15min 加入 $w(V) = 0.30\% \sim 0.50\%$, add 20min before tapping $w(V) = 0.30\% \sim 0.50\%$, 出钢前 20min 加入 $w(V) > 1\%$, reduction period, adjustment before tapping $w(V) > 1\%$, 还原期, 出钢前调整 Addition of molten steel, adjustment before tapping 熔化末加入, 出钢前调整	95~98
Ferrosilicon 硅铁		Before tapping 出钢前	95~98
Ferrotitanium 钛铁		Before tapping 出钢前	$w(Ti) \leqslant 0.15\%$, 30~50 $w(Ti) = 0.2\% \sim$ 0.8%, 50~65 $w(Ti) > 0.8\%$, 70~90
Aluminum ingot 铝锭	Smelting aluminum steel 冶炼铝钢	Before tapping 出钢前	$w(Al) = 1\%$, 70~80 $w(Al) = 2\%$, 80~85 $w(Al) = 3\%$, 85~90 $w(Al) > 5\%$, 90~95
Ferroboron 硼铁		Insert before tapping 出钢前插入 Add to package 包中加入	30~50
Rare earth 稀土		Insert before tapping 出钢前插入 Add to package 包中加入	20~40

3.4.4.2 Control and Adjustment of Composition of Molten Steel in Furnace
3.4.4.2 炉内钢液成分的控制与调整

Chemical composition has a great influence on the quality and properties of steel. A large number of scientific experiments and production practice show that the chemical composition of some steel grades should not only meet the requirements of technical conditions, but also be controlled in a more strict range, so as to meet the higher requirements for the quality and performance of the steel grades. For steel grades without special requirements, the composition is generally controlled according to the lower and middle limits, which can not only ensure the quality and performance requirements of steel, but also save alloy materials.

化学成分对钢的质量和性能均有很大的影响。大量的科学实验和生产实践表明，一些钢种的化学成分除应符合技术条件的规定外，还要控制在某一更加严格的范围内，才能满足于对该钢种质量和性能的更高要求。对于没有特殊要求的钢种，成分一般按中下限控制，这样既可保证钢的质量和性能的要求，又能节约合金材料。

In fact, the control of chemical composition of molten steel runs through the whole process of steel smelting: before charging the furnace, the steelmaker first checks the bill of materials, checks the charging amount of the furnace materials, the amount of carbon distribution and the quantity and composition of the alloy materials loaded, and immediately corrects any delay. After the results of full melting analysis are reported, the analysis composition of each element shall be checked against the specification composition of the steel. For the results of the content of residual elements exceeding or approaching the maximum allowable value, the verification shall be repeated, and then the smelting scheme shall be adjusted. It is not allowed to terminate the smelting of a furnace of steel easily because the content of residual elements is unclear or inaccurate. Experience tells us that the higher the ratio of impurity to iron is, the greater the fluctuation of residual element composition is, and the more careful we should be.

实际上，钢液化学成分的控制贯穿于一炉钢冶炼的始终：炉料入炉前，炼钢工首先核对料单，检查炉料的装入量、配碳量以及装入的合金材料的数量与成分是否合适，发现贻误应立即纠正。全熔分析结果报出后，要对照钢种的规格成分核对各元素的分析成分，对于残余元素的含量超过或接近最高许可值的结果，要反复核实验证，然后决定是否调整冶炼方案，绝不允许因残余元素含量不清、不准而轻易地终止一炉钢的冶炼。经验表明，杂铁比越高，残余元素成分波动越大，越要谨慎小心。

It is the key to control the final decarburization of molten steel at the end of oxidation. Under normal operation, the insufficient carbon in the steel should be absorbed from ferroalloy, electrode or reduction slag to make up for it, and the steel specification requirements can be met without carburization. Before tapping, if the carbon content is lower than the control specification, the pig iron with low S and P can also be used for carburizing, but the amount of carburizing should not be more than 0.05%, so as not to bring too much impurities into the steel and affect the quality of the steel. In addition, the S and P values brought in by pig iron and the content in steel shall not exceed the specification of steel. For the new lining with tar knot or carbide slag tapping and the new ladle made of tar brick, the liquid steel is easy to enter carbon in the tapping process, and the low carbon steel is easier to enter than the high carbon steel.

氧化末期的钢液成分控制终脱碳是关键。在操作正常的情况下，应该是钢中的不足之碳从铁合金、电极或还原渣中吸收而得到弥补，无须增碳即可满足钢种规格要求。出钢前，如果碳含量低于控制规格，还可用低S、P的增碳生铁进行增碳，但增碳量最好不大于0.05%，以免带入过多的杂质而影响钢的质量。除此之外，还需考虑生铁带入的S、P值加上钢中的也不许超过钢种规格的含量。对于焦油打结的新炉衬或电石渣出钢及用焦油砖砌制的新包，出钢过程钢液容易进碳，且低碳钢较比高碳钢进的更容易。

Phosphorus in steel is reduced by adding alloy and residual slag in the furnace, which increases slightly in the later stage of reduction. However, as long as the phosphorus content of full slag removal meets the specified slag removal conditions, it generally does not exceed the allowable value of steel specification. Under the condition of normal operation of basic slag, the sulfur in steel can be reduced to the required range smoothly. In addition, many can be removed during tapping. However, if it is found that the desulfurization is not good before tapping, proper measures shall be taken immediately to deal with it, and it is not allowed to decide the ultra sulfur tapping. The silicon in steel is usually obtained by the silicon powder added by indirect deoxidation, and the rest can be adjusted by silicon block before tapping. Of course, all of them can be adjusted with silicon block, which is of great practical significance for the rapid steelmaking of expanding precipitation deoxidization. Manganese in steel is usually obtained from the added manganese alloy in the process of deoxidization. The amount of residual manganese in steel should be considered when controlling. For high manganese steel or high silicon steel, when adjusting the addition of manganese or silicon before tapping, it is generally adjusted to the upper middle limit. This is due to the high content of manganese or silicon in the steel, and the oxidation loss is more in the process of tapping. For high silicon and high manganese steel, the affinity of silicon to oxygen is stronger than that of manganese, so silicon can be adjusted to the upper middle limit, while manganese can be adjusted to the middle limit. Before tapping, a large amount of aluminum or titanium can easily increase the silicon content in the steel. Therefore, for this kind of steel, the silicon content in the steel and the SiO_2 content in the slag should be controlled during the reduction operation. The chromium in the steel is obtained from the added ferrochromium except that the smelting by back blowing method or non oxidation method is carried into a part by special materials. For the content of nickel and aluminum elements, it should be seen that the two analysis results have little difference, and can be adjusted only when they are consistent with the material list. If it is inconsistent with the material list, the reason shall be found out or adjusted after multiple analysis and confirmation. For elements with high melting point that are not easy to oxidize, they are generally transferred into the lower and middle limits at one time. For the alloy elements that are easy to be reduced, they are generally adjusted to the lower limit or close to the lower limit, and the insufficient content is added before tapping. For the chromium steel which is sampled when smelting temperature is low, mixing is not good or adding time is short, the composition analysis is often high, but the actual content of liquid steel is not so high. During alloying, if more alloys are added in powder state, the composition analysis is easy to be low, and the actual content in the molten steel is indeed low, at this time, the composition adjustment should be higher. For the molten steel with low smelting temperature, bad mixing and high melting point elements, the composition analysis of high melting point elements is sometimes low and the actual content may not be low, so the composition adjustment should be careful. For continuous smelting of high alloy steel with high density and high melting point in multiple furnaces, the alloy elements with high density and high melting point in the first furnace should be controlled to the upper middle limit, and the other furnaces can be adjusted to the lower middle limit.

钢中的磷由于补加合金所带入的和残留炉中氧化渣中的磷还原，在还原后期略有增加，但只要全扒渣时的磷含量符合规定的扒渣条件，一般不会超过钢种规格的许可值。钢中的硫在碱性渣操作正常的情况下，一般均能很顺利地降到所要求的范围内。此外，在出钢过程中还能脱除许多。但在出钢前如果发现脱硫不好，应立即采取妥善的措施进行处理，切不可草率地决定超硫出钢。钢中的硅一般是通过间接脱氧所加入的硅粉获取，余者不足部分在出钢前可利用硅块补调。当然也可全部用硅块调入，这对扩大沉淀脱氧的快速炼钢来说，具有很大的实际意义。钢中的锰，一般是在脱氧过程中，从加入的锰合金上获取，控制时要考虑钢中的残余锰量。对于高锰钢或高硅钢，在出钢前调整补加锰或硅时，一般均调到偏中上限。这是由于钢中的锰或硅含量很高，在出钢过程中氧化损失较多。对于高硅高锰钢，因硅对氧的亲和力强于锰，所以硅调到中上限，而锰调到中限即可。出钢前大量地加铝或加钛易使钢中的硅含量增加，因此对于这类钢种，在还原操作过程中，应控制钢中的硅及渣中SiO_2的含量。钢中的铬除返吹法或不氧化法的冶炼由专用料带入一部分外，其余的均从加入的铬铁中获取。对于镍、铝元素含量应见到两个分析结果相差不大，且与料单配入相符才能进行调整。如与料单配入的不一致应查清原因或经多次分析确认无误再调整。对于高熔点不易氧化元素的成分一般一次调入中下限或中限。对于易还原的合金元素一般调到下限或接近下限，不足的含量在出钢前补加。对于冶炼温度偏低、搅拌又不好或加入的时间较短就取样的铬钢等，成分分析往往偏高，而钢液中实际含量却没有那么高。合金化时，如加入的合金呈粉末状态的较多，成分分析又容易偏低，而钢液中的实际含量也确实较低，这时成分的调整应偏高一些。对于冶炼温度偏低、搅拌又不好且含有较高的高熔点元素的钢液，高熔点元素的成分分析有时偏低而实际含量可能不低，这时成分的调整应要谨慎小心。对于连续冶炼多炉的密度大、高熔点的高合金钢，第一炉的密度大、高熔点的合金元素应往中上限控制，其余各炉一般调入中下限即可。

In addition, to control and adjust the composition of molten steel, it is necessary to know the composition of the last furnace steel, the influence of residual steel and residue on the composition and weight of the furnace steel. As for the steel grades washed by brush furnace, there are residual steel residues in the furnace or in the ladle, and the content of alloy elements is very high, some of them must be recovered. At this time, the control and adjustment of composition should be limited to the lower limit. The molten steel in the previous furnace is sometimes unclean and is engaged in the smelting of the next furnace steel. If the residual steel remaining in the furnace is more, it will inevitably affect the weight of the molten steel of the lower furnace steel. If the ingredients are adjusted in the furnace, if it is not considered, it will easily cause the chemical composition to be removed. Many years of production experience also shows that when the calculated composition is inconsistent with the analyzed composition, the reason should be found out and then adjusted, otherwise, the chemical composition will fluctuate. The control and adjustment of liquid steel composition is mainly carried out in the ladle for the operation of steel slag retention.

除此之外，钢液成分的控制与调整还要了解上一炉钢的成分、残钢与残渣对这炉钢成分及钢液重量的影响。对于刷炉洗包的钢种，由于炉内或包中留有前炉冶炼的残钢残渣，其中合金元素的含量又很高，势必有部分要被回收，这时成分的控制与调整应以进入下限即可。上一炉的钢液有时翻不净就从事下一炉钢的冶炼，如果炉中剩余的残钢较多，势必影响下炉钢的钢液重量，在炉中调整成分时，如不考虑极易引起化学成分的脱格。多年的生产经验还得出，当计算成分与分析成分不一致时，应查清原因后再调整，否则也会造成化学成分的波动。对于采用留钢留渣的操作，钢液成分的控制与调整主要是在包中进行。

In short, the control and adjustment of liquid steel composition must be comprehensively analyzed and comprehensively considered. A good EAF steelmaker should accumulate rich experience through long-term production practice in order to master this technology.

总之，钢液成分的控制与调整，必须全面分析、通盘考虑，一个好的电炉炼钢工应该通过长期的生产实践，积累丰富的经验，才能掌握这方面的技术。

3.4.4.3 Calculation of Liquid Steel Composition
3.4.4.3 钢液成分的计算

(1) Check of liquid steel weight. In actual production, due to factors such as inaccurate measurement, large fluctuation of charge quality or improper operation, the actual weight of liquid steel is easily inconsistent with the planned weight, which makes it difficult to control the chemical composition and pour the steel. Therefore, checking the actual weight of liquid steel can occasionally be encountered in production. Because the recovery rate of Ni and Mo elements in general alloy steel is relatively stable, the method of adjusting Ni and Mo can be used to check the weight of liquid steel. The check formula is as follows:

$$P\Delta b = P_0 \Delta b_0$$
$$P = P_0 \frac{\Delta b_0}{\Delta b}$$

Where　P——actual weight of liquid steel, kg;

P_0——original planned weight of molten steel, kg;

Δb——the amount of nickel or molybdenum increased analyzed in the furnace, %;

Δb_0——the amount of nickel or molybdenum increase calculated as P_0, %.

(1) 钢液重量的校核。在实际生产中，受计量不准或炉料质量波动较大或操作不当等因素的影响，极易出现钢液的实际重量与计划重量不符，而给化学成分的控制及钢的浇注造成困难。因此，校核钢液的实际重量在生产中偶尔也能遇到。因 Ni、Mo 元素在一般合金钢中的收得率比较稳定，所以可借调镍调钼的办法来校核钢液的重量，校核公式如下：

$$P\Delta b = P_0 \Delta b_0$$
$$P = P_0 \frac{\Delta b_0}{\Delta b}$$

式中　P——钢液的实际重量，kg;

P_0——原计划的钢液重量，kg；

Δb——炉中分析的增镍或增钼量，%；

Δb_0——按 P_0 计算的增镍或增钼量，%。

Example 3-1 The weight of the original planned molten steel is 20t, the content of Mo before adding Mo is 0.16%, the content of Mo calculated after adding Mo is 0.25%, and the actual analysis is 0.26%. Find the actual weight of liquid steel.

Solution: actual weight of molten steel = $20 \times \dfrac{(0.25 - 0.16)\%}{(0.26 - 0.16)\%} = 18(t)$ (3-3)

例 3-1 原计划钢液的重量为 20t，加钼前钼的含量（质量分数）为 0.16%，加钼后计算钼的含量为 0.25%，实际分析为 0.26%。求钢液的实际重量。

解： 钢液的实际重量 = $20 \times \dfrac{(0.25 - 0.16)\%}{(0.26 - 0.16)\%} = 18(t)$ (3-3)

It can be seen from Example 3-1 that the content of molybdenum in steel is only 0.01%, and the actual weight of liquid steel is 2t different from the original planned direct quantity. However, the deviation of ±(0.01% ~ 0.03%) is easy to appear in chemical analysis, so it is difficult to accurately check and judge the actual weight of liquid steel. Therefore, the formula (3-3) is only applicable to theoretical calculation, while in the actual smelting process, the formula (3-4) is generally used for checking the weight of liquid steel:

$$P = \frac{GC}{\Delta b} \quad (3-4)$$

Where P——actual weight of liquid steel, kg；

G——supplement amount of nickel or ferromolybdenum, kg；

C——composition of nickel or ferromolybdenum, %；

Δb——the amount of nickel or molybdenum increased analyzed in the furnace, %.

由例 3-1 可以看出，钢中钼的含量（质量分数）仅差 0.01%，钢液的实际重量就与原计划直量相差 2t，而化学分析往往容易出现±(0.01% ~ 0.03%) 的偏差，这样就很难准确地校核判断钢液的实际重量。因此，式（3-3）只适用于理论上的计算，而在实际冶炼过程中，钢液重量的校核一般均采用式（3-4）计算：

$$P = \frac{GC}{\Delta b} \quad (3-4)$$

式中 P——钢液的实际重量，kg；

G——镍或钼铁的补加量，kg；

C——镍或钼铁的成分，%；

Δb——炉中分析的增镍或增钼量，%。

Example 3-2 Add 15kg of ferromolybdenum to the furnace, and the molybdenum content in the molten steel increases from 0.20% to 0.25%. It is known that the molybdenum content in ferromolybdenum is 60%. Calculate the actual weight of the molten steel in the furnace.

Solution: actual weight of melten steel in the furnace = $\dfrac{15 \times 60\%}{0.25\% - 0.20\%} = 18000(\text{kg})$

例 3-2 往炉中加入钼铁 15kg，钢液中的钼含量（质量分数）由 0.20% 增到 0.25%，已知钼铁中钼的成分为60%,求炉中钢液的实际重量。

解: 炉中钢液的实际重量 $= \dfrac{15 \times 60\%}{0.25\% - 0.20\%} = 18000(\text{kg})$

Example 3-3 For smelting 20CrNiA steel, due to the temporary failure of the electronic scale, the loaded steel material was not weighed, so the loader estimated the loading. To find the weight of molten steel in the furnace.

Solution: add 100kg nickel plate into the furnace, the nickel content in the molten steel increases from 0.90% to 1.20%, and the known composition of nickel plate is 99%, then:

$$\text{weight of melten steel in the furnace} = \dfrac{100 \times 99\%}{1.20\% - 0.90\%} = 33000(\text{kg})$$

例 3-3 冶炼20CrNiA钢,因电子秤临时出故障,装入的钢铁料没经称量,由装料工估算装料。求炉中钢液重量。

解: 往炉中加入镍板100kg,钢液中的镍含量由0.90%增到1.20%,已知镍板的成分为99%,则:

$$\text{炉中钢液重量} = \dfrac{100 \times 99\%}{1.20\% - 0.90\%} = 33000(\text{kg})$$

For liquid steel without nickel or molybdenum, the weight verification mainly depends on experience. As manganese is greatly affected by smelting temperature and oxygen and sulfur content in steel, the use of manganese elements to check the weight of molten steel can only be carried out at the end of reduction, but the accuracy is poor in the oxidation process or the early reduction.

对于不含镍或钼的钢液,重量的校核主要凭借经验。由于锰受冶炼温度及钢中的氧、硫含量的影响较大,因此利用锰元素来校核钢液的重量,只能在还原末期进行,而在氧化过程中或还原初期的准确性较差。

(2) The calculation formula for the addition of a single alloy element is:

$$P = KQ$$

Where P——weight of liquid steel, kg;

Q——loading capacity, kg;

K——comprehensive recovery rate of charge, %.

(2) 单一合金元素的加入计算,计算公式为:

$$P = KQ$$

式中 P——钢液重量, kg;

Q——装料量, kg;

K——炉料的综合收得率, %。

$$a = \dfrac{bP + fcG}{P + G} \tag{3-5}$$

Where G——ferroalloy content, kg;

a——alloy element control specification composition, %;

b——analysis composition of elements in the furnace, %;

c——element composition in ferroalloy, %;

f——yield of alloy element,%.

$$a = \frac{bP + fcG}{P + G} \tag{3-5}$$

式中 G——铁合金加入量，kg；
a——合金元素控制规格成分，%；
b——炉中元素的分析成分，%；
c——铁合金中的元素成分，%；
f——合金元素的收得率，%。

When the specification content of the elements in the steel is not high, the effect of the alloy content on the total weight of the liquid steel can be ignored, that is, the G of the denominator in the formula (3-5) is omitted, and the calculation formula of the alloy element addition can be simplified as follows:

$$G = P\frac{a-b}{fc} \tag{3-6}$$

当钢中元素的规格含量不高时，合金用量对钢液总重量的影响可忽略不计，即式（3-5）中分母的 G 略去，合金元素加入的计算公式可简化为：

$$G = P\frac{a-b}{fc} \tag{3-6}$$

The formula (3-6) is applicable to the addition calculation of alloy elements in carbon steel or low alloy steel, i. e. the steel grades with element content less than 3% or the sum of other alloy element content less than 3.5%. The higher the content of elements in steel, the larger the calculation error, but it can not be out of specification.

式（3-6）适用于碳素钢或低合金钢合金元素的加入计算，即指单元合金元素含量小于3%或加上其他合金元素含量的总和小于3.5%的钢种。而钢中元素的含量越高，计算误差越偏大，但也不能脱出规格。

Example 3-4 Smelting 38CrMoAlA Steel. It is known that the charging capacity is 20t, the burning loss of the charge is 4%, the analyzed aluminum content in the furnace is 0.05%, the control specification composition of aluminum is 0.95%, the aluminum composition in the aluminum ingot is 98%, and the aluminum recovery rate is 75%. How much aluminum ingot to add?

Solution： amount of aluminum ingot added = $2000 \times (100-4)\% \times \frac{(0.95-0.05)\%}{98 \times 75\%} = 235.10(kg)$

例 3-4 冶炼 38CrMoAlA 钢。已知装料量为20t，炉料烧损为4%，炉中分析铝含量为0.05%，铝的控制规格成分为0.95%，铝锭中铝的成分为98%，铝的收得率为75%。求铝锭加入量。

解： 铝锭加入量 = $2000 \times (100-4)\% \times \frac{(0.95-0.05)\%}{98 \times 75\%} = 235.10(kg)$

Exercises

(1) What is the task of restore period?
(2) What are the deoxidization methods for EAF steelmaking?

(3) What are the strengthening measures for desulfurization operation in reduction period?
(4) What are the principles of adding alloy elements?

思考题
(1) 还原期的任务是什么？
(2) 电炉炼钢的脱氧方法有哪些？
(3) 还原期脱硫操作的强化措施有哪些？
(4) 合金元素加入的原则有哪些？

Task 3.5　Tapping Operation
任务 3.5　出钢操作

Mission objectives
任务目标

(1) Understand the tapping conditions of electric furnace.
(1) 了解电炉的出钢条件。
(2) Master the tapping method of electric furnace.
(2) 掌握电炉出钢方式。
(3) Master the operation of electric furnace tapping.
(3) 掌握电炉出钢操作。

3.5.1　Tapping Conditions of Electric Furnace
3.5.1　电炉的出钢条件

Tapping is the last operation of smelting in front of the furnace, but it can only be done with tapping conditions, otherwise it will affect the quality and output of steel. The tapping of traditional electric furnace does not include the operation of leaving steel and slag. The tapping conditions are as follows:

出钢是炉前冶炼的最后一项操作，但必须具备出钢条件才能出钢，否则将会影响钢的质量和产量。传统电炉的出钢不包括留钢留渣操作，出钢条件如下：

(1) All chemical components enter the control specification. Before the tapping, the chemical composition of the liquid steel must be all adjusted to the specifications before the furnace is adjusted.

(1) 化学成分全部进入控制规格。出钢前，钢液的化学成分凡属炉前调整的必须全部进入规格，没进入控制规格或没有满足合同要求的不准出钢。

(2) The tapping temperature meets the requirements. Proper tapping temperature is not only

the key to ensure the quality of molten steel, but also one of the first conditions to ensure the smooth operation of pouring. Therefore, the tapping temperature of the molten steel must meet the technological requirements of the steel-making process. It is not only necessary to avoid the high temperature molten steel, but also not the low temperature steel that cannot be poured or barely maintained.

（2）出钢温度合乎要求。合适的出钢温度不仅是保证钢液质量的关键，而且也是保证浇注操作顺利进行的首要条件之一。因此，钢液的出钢温度必须满足所炼钢种的工艺要求，既要避免出温度过高的高温钢液，也不许出不能浇注或勉强维持浇注的低温钢。

(3) Liquid steel deoxidation must be good. In order to reduce the sundries in the center of the steel, improve the purity of the steel as much as possible, and prevent the rise of the cap mouth, bubbles, hairlines and other defects, the liquid steel with poor deoxidation of the electric furnace steel cannot be used for pouring. Therefore, for the liquid steel without deoxidizing means outside the furnace, it is required to deoxidize well in the furnace, otherwise the steel can not be tapped.

（3）钢液脱氧必须良好。为了降低钢中央杂物，尽量提高钢的纯洁度，防止帽口上涨、气泡和发纹等缺陷的产生，电炉钢脱氧不良的钢液不能用来浇注。因此，对于没有炉外脱氧手段的钢液，要求炉中必须脱氧良好，否则不能出钢。

(4) The fluidity and basicity of slag should be suitable. Before tapping, the fluidity and basicity of slag should be good. Because the viscous slag will lead to steel first and then slag, or only the molten steel will not be slag or the slag will be very little, it is easy to make the temperature of molten steel drop sharply, which will affect the control of normal sedation and temperature injection speed. In addition, the viscous slag is easy to jam the plug rod in the ladle and bring difficulties to the pouring work, and the sticky slag mixed in the steel is not easy to float up and affect the internal quality of the steel; the too thin slag will also make the molten steel cool down quickly and easily erode the ladle wall and plug rod, the lighter will reduce the service life of the ladle, the heavier will erode the plug rod and affect the smooth operation of the pouring steel. Proper basicity can reduce the oxygen content in steel and remove the sulfur in steel. In addition, the basicity can also change the activity of SiO_2 in slag, which is very important for the smelting of high silicon steel and high aluminum steel. Therefore, it is necessary to make the slag have proper basicity according to the specific situation before tapping.

（4）熔渣的流动性和碱度要合适。出钢前，熔渣的流动性要好，碱度也要合适。因黏稠的熔渣在出钢时会造成先出钢后出渣，或只出钢液不出渣或熔渣出得很少，易使钢液温度急剧下降，影响正常的镇静与注温注速的控制。此外，黏稠的熔渣在包中极易卡住塞杆而给浇注工作带来困难，而且混在钢中的粘渣不易上浮而影响钢的内在质量；过稀的熔渣也会使钢液降温快，并极易侵蚀包壁和塞杆，轻者降低钢包的使用寿命，重者蚀断塞杆影响浇钢操作的顺利进行。合适的碱度有利于降低钢中的氧含量，也能较好地去除钢中的硫。此外，碱度的高低还能改变渣中 SiO_2 的活度，这一点对于高硅钢和高铝钢的冶炼极为重要。因此，出钢前应针对具体情况，使熔渣要有合适的碱度是十分必要的。

(5) Slag quantity and color shall be normal. Before tapping, the slag quantity and slag color in the furnace shall be normal. If the amount of slag in the furnace is small or the amount of slag turned into the ladle is not enough, it is easy to cause the liquid steel to be exposed, not only to inhale seriously, but also to cool down or stick to the ladle bottom. What's more, too much sticking to the ladle bottom will make the pouring amount insufficient. Too much slag will not only increase the consumption of slagging materials, but also affect the accurate judgment of steel water.

(5) 渣量和渣色要正常。出钢前，炉中的渣量和渣色要正常。如果炉中渣量少或翻入包中的渣量不够，易造成钢液裸露，不仅吸气严重，而且也极易降温或出现粘包底现象，更有甚者，因粘包底过多会使浇注量不足。渣量过大既会增加造渣材料的消耗，又影响钢水量的准确判断。

When tapping, white slag or white slag are generally required, and carbide slag or yellow slag are the most taboo. Due to the good wetting of carbide slag, slag steel is not easy to separate, which affects the floatation and removal of non-metallic inclusions. However, the yellow slag shows high content of (FeO) or (MnO). In addition, the ladle and pouring system shall be prepared in advance to ensure the normal operation of the equipment, smooth steel outlet, flat steel outlet, clean and dry steel outlet pit, clean furnace cover and steel outlet, etc., which are also necessary preparations before steel tapping.

出钢时，一般要求白渣或花白渣，最忌讳电石渣或黄渣。因电石渣润湿较好，渣钢不易分离，影响非金属夹杂物的上浮与去除。而黄渣表明（FeO）或（MnO）含量较高。除此之外，提前备好钢包和浇注系统，保证设备运转正常，并使出钢口畅通、出钢槽平整、出钢坑清洁干燥、炉盖和出钢槽吹扫干净等，也是出钢前必须要做好的准备工作。

In order to make the tapping operation rule-based, a factory summed up the regulation of 'five no tapping' based on the accumulated experience for many years, which is worthy of reference. The specific contents are as follows:

(1) In tapping analysis, the chemical composition of the two samples is too different, so it is not allowed to tap without finding out the cause and treatment;

(2) When the calculated composition is inconsistent with the analyzed composition, no tapping is allowed without finding out the cause;

(3) The adjustment of chemical composition is not allowed to tap steel according to the calculation or analysis without entering the plant control, internal control specifications or meeting the contract requirements;

(4) If the slag and temperature do not meet the requirements, no tapping is allowed;

(5) No tapping is allowed without the approval of on duty and supervision personnel.

为使出钢操作有章可循，某厂根据多年积累的经验，总结出"五不出钢"的规定很值得借鉴，具体内容如下：

(1) 出钢分析两个样的化学成分相差太大，没有查清原因及未经处理不准出钢；

(2) 计算成分和分析成分不一致时，没查出原因不准出钢；

(3) 化学成分的调整按计算或分析未进入厂控、内控规格或没有满足合同要求也不准出钢；

（4）渣子和温度未达到要求不准出钢；

（5）不经值班和监督人员的批准不准出钢。

3.5.2 Tapping Method of Electric Furnace
3.5.2 电炉的出钢方式

There are three ways of tapping in traditional electric furnaces:

(1) Deep pit, large mouth spitting, slag steel mixed flushing. The advantage of this tapping method is that the molten steel can be well protected by slag, which not only reduces the temperature drop of molten steel, but also controls the secondary oxidation and suction of molten steel. In addition, the non-metallic inclusions suspended in the steel can also be fully washed by slag, which is conducive to floatation and removal, and the large orifice spray can further deoxidize and desulfurize. Therefore, this kind of tapping method is quite common.

传统电炉的出钢方式有 3 种：

（1）深坑、大口喷吐、渣钢混冲。这种出钢方式的优点是钢液能得到熔渣较好的保护，既减少了钢液的降温，又可控制钢液的二次氧化与吸气。此外，钢中悬浮的非金属夹杂物也能得到熔渣充分的洗涤，利于上浮与去除，大口喷吐还可进一步脱氧与脱硫。因此，该种出钢方式比较常见。

(2) First, the liquid steel is produced and then the slag is produced, which is also called slag retaining tapping. The advantages of this tapping method are mainly to improve the yield of alloying elements in some ladle and to stabilize the chemical composition of the molten steel. The disadvantages of this tapping method are that the molten steel has fast cooling speed and poor deoxidization and desulfurization capacity.

（2）先出钢液后出渣，又称挡渣出钢。这种出钢方式的优点主要是能提高某些包中合金化元素的收得率和稳定钢液的化学成分，缺点是钢液降温快，包中脱氧、脱硫能力差。

(3) It is the combination of the above two tapping methods. Generally, the slag is first mixed with steel, and the large mouth is sprayed for a while, then the slag is blocked for tapping, and finally a large amount of slag is discharged; or the slag is blocked for tapping, and then the slag is mixed with steel, and the large mouth is sprayed, and the steelmaker can choose at will according to the needs.

（3）上述两种出钢方式的结合。一般先渣钢混中，大口喷吐一阵，然后再挡渣出钢，最后大量出渣；也可先挡渣出钢，然后渣钢混中，大口喷吐，炼钢工可根据需要选择。

3.5.3 Tapping Operation of Electric Furnace
3.5.3 电炉的出钢操作

Due to the large tilting angle of the traditional electric furnace, the electrode should be raised properly before tapping, especially the electrode near the tapping hole. It should be noted that the slag liquid with too high electrode temperature drops quickly and is not easy to flow out. For example, the operation of the crane is affected by the use of crane ladle tapping; the carbon of the liquid steel is easily increased by too low electrode. Then cut off the power supply. It is strictly prohibited to tap steel with electricity to prevent short circuit. The tap hole shall be large. In order to ensure the purity of the liquid steel, the plug shall not be pushed into the furnace. When tapping, the time should be shortened as much as possible to reduce the secondary oxidation and gas absorption of the liquid steel. In addition, it is also strictly prevented from small flow, scattered flow and choking. Because the fine flow and scattered flow tapping not only can make the molten steel cool down quickly, but also the dynamic condition of tapping is not good, which will affect the deoxidization, desulfurization and the floatation and removal of non-metallic inclusions in the ladle; the choking out steel capacity will cause the slag steel to overflow and splash, thus increasing the cleaning amount after tapping. In addition, in the process of tapping, it is also necessary to avoid damaging the tapping groove or ladle, and strictly prevent the steel flow from impacting the stopper rod or ladle wall.

传统电炉的出钢由于炉体倾动角度较大，因此出钢前首先要适当升高电极，特别是靠近出钢口处的电极尤要注意，电极过高渣液降温快不易流出，如采用吊车吊包出钢，影响吊车的运转；电极过低易使钢液增碳。然后切断电源，严禁带电出钢，以防短路。出钢口要掏大，为保证钢液的纯洁，堵塞物不准推进炉内。出钢时还要尽量缩短时间，以减少钢液的二次氧化与吸气。另外，也严防细流、散流、呛流出钢。因细流、散流出钢不仅能使钢液降温快，出钢动力学条件也不好，进而影响包中的脱氧、脱硫及非金属夹杂物的上浮与去除；呛流出钢容量造成渣钢横溢飞溅，从而增加了出钢后的清理量。除此之外，出钢过程中还要避免撞坏出钢槽或钢包、严防钢流冲击塞棒或包壁。

According to the requirements of smelting process, some steel grades need to be deoxidized or chemical composition adjusted during tapping, such as adding aluminum block, silicon calcium block or adding ferroboron, rare earth elements and other alloys. At this time, appropriate timing and tapping method should be selected. In the process of tapping, the personnel on duty, steelmakers and pouring workers shall pay attention to observe the tapping temperature and determine the sedation time based on other conditions. After discharging the steel, add some carbonized rice husk or straw ash or carbon powder (but not economical) to the slag surface around the stopper rod, and then smash the slag cover to make it fluffy, so as to prevent the stopper rod from being blocked by slag during the process of steel liquid sedation, which may affect the opening and closing or cause bending. For high temperature molten steel or low alkalinity slag, lime should be added around the plug rod to reduce the slag temperature or make it thickened.

根据冶炼工艺要求，有的钢种需在出钢过程中进行终脱氧或调整化学成分，如加入铝块、硅钙块或加入硼铁、稀土元素及其他合金等，这时要选择合适的时机和出钢方式。在出钢过程中，值班人员和炼钢工及浇注工应注意观察出钢温度，综合其他情况确定镇静时间。出完钢后，应往塞杆周围的渣面上加些碳化稻壳或草灰，也可加些炭粉（但不够经济），然后打砸这部分渣盖，使其蓬松，预防钢液在镇静过程中，渣子将塞杆卡住而影响启闭或造成弯曲。对于高温钢液或碱度低的稀渣，应往塞杆周围加些石灰等物，以降低这部分渣温或使其变稠。

3.5.4　Experience Judgment of Tapping Temperature
3.5.4　出钢温度的经验判断

At present, the continuous temperature measurement of ladle is simple and accurate, which has provided important temperature parameters for steel pouring. However, people still don't give up the experience judgment of tapping temperature completely, that is to say, the experience judgment of steel temperature still has great reference value. The commonly used methods are as follows:

目前，钢包的连续测温既简便又准确，已为钢的浇注提供了重要的温度参数。尽管如此，人们还是没有完全放弃出钢温度的经验判断，也就是说出钢温度的经验判断仍然具有较大的参考价值，常用的方法主要有以下几种：

(1) Visually measure the tapping temperature of the liquid steel. Generally, the tapping temperature can be judged by observing the color of the steel flow 100~200mm away from the end of the tapping trough during the tapping process and when there is slag cover.

1) The liquid steel is dark red and below 1550℃;
2) The liquid steel is bright red, about 1600℃;
3) The liquid steel is blue white, about 1620℃;
4) The liquid steel is blue white, white smoke is seen at the upper part of the tapping tank, about 1630℃;
5) The liquid steel is white, and thick white smoke is emitted from the upper part of the tapping tank, about 1650℃;
6) The molten steel is dazzling, and the white smoke rolling on the upper part of the steel chute is above 1670℃.

(1) 目测钢液的出钢温度。通常，出钢温度可在出钢过程并在有熔渣遮盖时，观察距出钢槽端部外 100~200mm 处的钢流颜色进行判断：

1) 钢液呈暗红色，约在 1550℃ 以下；
2) 钢液呈亮红色，约为 1600℃；
3) 钢液呈青白色，约为 1620℃；
4) 钢液呈青白色，出钢槽上部见白烟，约为 1630℃；
5) 钢液呈白色，出钢槽上部冒浓浓白烟，约为 1650℃；
6) 钢液自炽耀眼，出钢槽上部白烟滚滚，约在 1670℃ 以上。

In the process of tapping, the smoke produced by a large amount of manganese oxidation affects the line of sight. It is difficult to judge the true color without careful discrimination. Due to the high content of Cr, Ni, W, Mo and other elements in high alloy steel, it is easy to make people have illusion in the process of tapping. In addition, the visual inspection of the tapping temperature is also affected by the temperature of the lining slag. The general rule is: the higher the temperature of the lining slag is, the lower the temperature of the liquid steel is, such as the tapping temperature of the post heating; and the lower the temperature of the lining slag is, the higher the temperature of the liquid steel is, such as the tapping temperature of the slag change before tapping or the sharp cooling of the power cut.

  锰含量较高的钢种，在出钢过程中，因锰元素的大量氧化所产生的烟雾，影响视线，如不仔细分辨难以判断真实的颜色。高合金钢中由于含有较高的 Cr、Ni、W、Mo 等元素，在出钢过程中，也极易使人们产生错觉。此外，就是出钢温度的目测还受衬托熔渣温度的影响，一般规律为：衬托的熔渣温度越高，目测钢液的温度越显得偏低，如后升温的出钢温度；而衬托的熔渣温度越低，目测钢液的温度就越显得偏高，如出钢前的换渣或停电急剧降温等的出钢温度。

  For a large furnace with a large load, the color of the molten steel may change several times during the tapping process due to the poor mixing of the population. At this time, different temperatures before and after should be converted. Do not use one or a while of molten steel temperature to represent the tapping temperature of the whole furnace.

  对于装入量较多的大炉子，由于入口搅拌不好，在出钢过程中，钢液的颜色可能变换几次。这时应对前后不同的温度进行折算，不要用一时或一阵的钢液温度来代表全炉的出钢温度。

  （2）Based on the change of slag in ladle, the tapping temperature of molten steel is estimated. In the process of tapping, when a certain amount of molten steel is poured out, the slag in the ladle suddenly thickens and thins, indicating that the tapping temperature is high. If the slag has the tendency of breaking the plug bar, the tapping temperature is higher. When the molten steel is discharged, the change of slag's thickness is not obvious, which indicates that the tapping temperature is general.

  （2）利用包中熔渣的变化大概估计钢液的出钢温度。在出钢过程中，当钢液翻出一定量后，包中熔渣突然由稠变稀，说明出钢温度较高。如熔渣大有蚀断塞棒的趋势，说明出钢温度更高。当钢液出完后，包中熔渣稀稠的变化不大，说明出钢温度一般。

  （3）The molten circle between ladle wall and slag steel is used to judge the tapping temperature of molten steel. The higher the temperature is, the more serious the erosion is. The melting circle between the ladle wall and the slag cover reflects the degree of the erosion. Therefore, the state of the melting circle can roughly represent the temperature of tapping: if there is no melting circle between the ladle wall and the slag cover, the tapping temperature is generally less than 1600℃; if there is only bubbling between the ladle wall and the slag cover, the tapping temperature is about 1610℃; if there is a melting circle between the ladle wall and the slag cover, the tapping temperature is about 1620℃ above, the wider the melting circle, the higher the tapping temperature; for example, the boiling of the melting circle, the tapping temperature is above 1650℃. Of course, this kind of melting circle is related to the lining material and basicity of slag. The above conclusion is mainly applicable to the wall of clay brick and slag with medium basicity.

  （3）利用包壁与渣钢间的熔融圈来判断钢液的出钢温度。还原渣对包壁的耐火材料有侵蚀，温度越高，这种侵蚀越严重，而包壁与渣盖间的熔融圈反映了这种侵蚀的程度。因此，熔融圈的状态也就能粗略表示出钢温度的高低：如包壁与渣盖间没有熔融圈，出钢温度一般均小于1600℃；如包壁与渣盖间仅是冒泡，出钢温度约为1610℃；如包壁与渣盖间出现熔融圈，出钢温度约有1620℃以上，熔融圈越宽，出钢温度越高；如熔融圈翻滚沸腾，出钢温度约在1650℃以上。当然，这种熔融圈与包衬的材质及熔渣的碱度有关，上述的结论主要适用于黏土砖的包壁和中等碱度的熔渣。

Exercises

（1） What is the tapping condition of electric furnace?
（2） What are the ways of tapping?

思考题

（1） 电炉的出钢条件是什么？
（2） 出钢方式有哪些？

Project 4 Masonry and Maintenance of EAF Lining
项目 4 电炉炉衬的砌筑和维护

Task 4.1 Main Properties and Classification of Refractories
任务 4.1 耐火材料的主要性能和分类

Mission objectives
任务目标

(1) Understand the types of refractory materials.
(1) 了解耐火材料的种类。
(2) Master the performance indexes of refractory materials.
(2) 掌握耐火材料的性能指标。

Refractory is a kind of solid material which can resist the action of high temperature (above 1580℃). Refractory is an indispensable lining material for all industrial furnaces, and its application range is extremely wide. Among them, metallurgical industry accounts for about 70% of the total refractory production. However, at present, there is not a kind of refractory that can fully meet the requirements of service performance, even if the performance of the same refractory under different service conditions is not the same. Therefore, in order to use refractories reasonably, it is necessary to know the performance and working conditions of refractories.

耐火材料是一种能抵抗高温（1580℃以上）作用的固体材料。耐火材料是所有工业用炉不可缺少的内衬材料，其使用范围极其广泛，其中冶金工业用量最大，约占耐火材料生产总量的70%。但是，目前尚没有一种耐火材料能够完全满足使用性能的要求，即使同一耐火材料在不同的使用条件下所表现的性能也不相同。因此，为了合理使用耐火材料，必须了解耐火材料的性能和使用的工作条件。

4.1.1 Main Performance Indexes of Refractories
4.1.1 耐火材料的主要性能指标

The main performance indexes of refractories include:

(1) Refractoriness. Refractoriness is defined as the resistance of refractories to high tempera-

ture without melting. Refractory does not have a fixed melting point, so refractoriness actually refers to the temperature at which the refractory is softened to a certain degree. Refractoriness is an important index of refractoriness. The refractoriness of selected refractoriness should be higher than its maximum operating temperature.

耐火材料的主要性能指标包括：

（1）耐火度。耐火材料抵抗高温作用而不熔化的性能称为耐火度。耐火材料没有固定的熔点，所以耐火度实际上是指耐火材料软化到一定程度时的温度。耐火度是耐火材料的重要指标，选用耐火材料的耐火度，应高于其最高使用温度。

（2）Thermal stability. Thermal stability refers to the ability of refractory materials to resist cracking and damage in the case of rapid temperature change, and the ability to resist cracking or crushing in use.

（2）热稳定性。耐火材料承受温度急剧变化而不开裂、不损坏的能力，以及在使用中抵抗破裂或破碎的能力，称为热稳定性。

（3）Slag resistance. The ability of refractories to resist slag erosion at high temperature is called slag resistance.

（3）抗渣性。耐火材料在高温下抵抗炉渣侵蚀的能力称为抗渣性。

（4）Volume stability. The ability of refractories to resist volume change when they are heated for a long time under high temperature is called volume stability, also known as reheat shrinkage or expansion.

（4）体积稳定性。耐火材料在高温下长期受热时抵抗体积变化的能力称为体积稳定性，也称重烧收缩或膨胀。

（5）Load softening temperature. The temperature at which a certain amount of deformation is caused by a static load of $2kg/cm^2$ of refractory under high temperature, which is called load softening temperature. In production, the actual load of the material is usually less than $2kg/cm^2$, and the one side is more heated, the colder side bears most of the load. Therefore, the maximum service temperature of the refractory is usually higher than its load softening temperature.

（5）荷重软化温度。耐火材料在高温时，每平方厘米承受2kg静负荷作用下，引起一定数量变形的温度，称为荷重软化温度。在生产中，材料实际负荷常小于$2kg/cm^2$，同时是单面受热的多，较冷的一面承受大部分负荷，因此耐火材料的最高使用温度常较其荷重软化温度为高。

（6）Bulk density. The ratio of the mass to volume of the refractories dried at 110℃ is called bulk density, and the unit is g/cm^3.

（6）体积密度。耐火材料在110℃下干燥后的质量与体积之比称为体积密度，单位为g/cm^3。

（7）True density. The ratio of the mass of refractories dried at 110℃ to the true volume, called true density, and the unit is g/cm^3. True volume refers to the difference between the total volume of the sample and the volume occupied by the pores in the sample.

（7）真密度。耐火材料在110℃下干燥后的质量同真体积之比，称真密度，单位为 g/cm³。真体积指试样总体积与试样中孔隙所占的体积之差。

(8) Porosity. Porosity includes apparent porosity and real porosity. Apparent porosity: the ratio of the volume to the total volume of the pores in contact with the atmosphere (open pores). True porosity: the ratio of the volume of all pores (including the volume of open and closed pores) of refractory to the total volume. Both porosity are expressed as percentage.

(8) 气孔率。气孔率包括显气孔率和真气孔率。显气孔率：耐火材料与大气相通的孔隙（开口孔隙）的体积与总体积之比。真气孔率：耐火材料全部孔隙的体积（包括开口和闭口孔隙的体积）与总体积之比。两种气孔率都用百分比表示。

(9) Compressive strength at room temperature. The capacity of refractories to bear load per square centimeter at room temperature is called compressive strength at room temperature, unit is kgf/cm² (0.1MPa).

(9) 常温耐压强度。耐火材料在常温下每平方厘米承受负荷的能力称常温耐压强度，单位为 kgf/cm² (0.1MPa)。

In addition to the above performance indexes, refractories also have some important indexes such as thermal conductivity, conductivity, plasticity, permeability and water absorption. The shape and size of refractories also have a great influence on the practical application of refractories, which directly affects the brick joints in masonry. The visual inspection items of refractories include: dimensional tolerance, angle missing, edge missing, distortion, crack, melting hole and slag corrosion, etc. The dimensional tolerance of general products shall not exceed ± 3%.

除上述性能指标外，耐火材料还有导热性、导电性、可塑性、透气性及吸水率等一些重要的指标。耐火材料的外形和尺寸，对于耐火材料的实际应用也具有很大的影响，它直接影响到砌筑时的砖缝。耐火材料的外观检查项目有尺寸公差、缺角、缺棱、扭曲、裂纹、熔洞和渣蚀等。一般制品的尺寸公差不得超过±3%。

4.1.2 Classification of Refractories
4.1.2 耐火材料的分类

There are many kinds of refractories. According to different purposes and requirements, there are many classification methods. The commonly used classification methods are as follows:

耐火材料的种类繁多，根据不同的使用目的和要求，有许多分类方法，常用的几种分类方法如下：

(1) Classified by fire resistance: ordinary refractory, the refractoriness is 1580~1770℃; advanced refractory, the refractoriness is 1770~2000℃; superior refractory, the refractoriness is 2000~3000℃; super refractory, the refractoriness is more than 3000℃.

(1) 按耐火度分类有：普通耐火材料，耐火度为 1580~1770℃；高级耐火材料，耐火度为 1770~2000℃；特级耐火材料，耐火度为 2000~3000℃；超特级耐火材料，耐火度为 3000℃以上。

(2) According to chemical properties: acid refractory includes quartz (silica), silica brick; semi-acid refractory includes semi-silicon brick; neutral refractory includes chrome

brick, clay brick, high alumina brick, clay refractory, etc; basic refractory includes magnesia brick, chrome magnesia brick, magnesia aluminum brick, dolomite brick, magnesia, dolomite and magnesia refractory mortar, etc.

（2）按化学性质分类有：酸性耐火材料包括石英（硅石）、硅砖；半酸性耐火材料包括半硅砖；中性耐火材料包括铬砖、黏土砖、高铝砖、粘土质耐火泥等；碱性耐火材料包括镁砖、铬镁砖、镁铝砖、白云石砖、镁砂、白云石及镁质耐火泥等。

（3）According to the product size classification: standard brick includes rectangular brick, longitudinal wedge brick, horizontal wedge brick, the most common standard brick size of 230mm×113mm×65mm; universal type brick; heterotypic brick; special shaped brick.

（3）按制品外形尺寸分类有：标准型砖：直角形砖、纵楔形砖、横楔形砖，最常见的标准型砖的尺寸230mm×113mm×65mm；普型砖；异型砖；特异型砖。

In addition, refractory materials can also be classified by use. Such as electric roof brick, steel bucket brick; according to the different production process of refractory materials can be divided into fired brick, not fired brick, molten brick, etc.

此外，耐火材料还可按用途分类。如电炉炉盖用砖、盛钢桶用砖等；按耐火材料制作工艺不同还可分为烧成砖、不烧砖、熔铸砖等。

Exercises
(1) What are the main performance indicators of refractory materials?
(2) What types of refractory materials are classified according to their chemical properties?
(3) What types of refractory materials are classified according to the refractoriness?

思考题
（1）耐火材料的主要性能指标有哪些？
（2）耐火材料按照化学性质分成哪几类？
（3）耐火材料按照耐火度分成哪几类？

Task 4.2　Refractories for EAF
任务4.2　电弧炉用耐火材料

Mission objectives
任务目标

(1) Master the refractory materials used in various parts of the EAF.
（1）掌握电炉各部位使用的耐火材料。
(2) Master the general requirements of refractory materials for EAF steelmaking.
（2）掌握电炉炼钢对耐火材料的一般要求。

4.2.1 General Requirements for Refractories in EAF Steelmaking
4.2.1 电炉炼钢对耐火材料的一般要求

The general requirements for refractories in EAF steelmaking are as follows:
电炉炼钢对耐火材料的一般要求有：

(1) High refractoriness. Refractories must have high refractoriness, because the arc temperature is above 4000℃, and the steel-making temperature is usually 1500~1750℃, sometimes even up to 2000℃.

（1）高耐火度。电弧温度在4000℃以上，炼钢温度常在1500~1750℃，有时甚至高达2000℃，因此，要求耐火材料必须有高的耐火度。

(2) High load softening temperature. Refractories must have high load softening temperature, because EAF steel-making process is working under high temperature load conditions and the furnace lining is subject to the erosion of molten steel.

（2）高荷重软化温度。电炉炼钢过程是在高温载荷条件下工作的，并且炉体要经受钢水的冲刷，因此耐火材料必须有高的荷重软化温度。

(3) Good thermal stability. The temperature of EAF steel-making changes rapidly within a few minutes from tapping to charging, and the temperature suddenly drops from about 1600℃ to below 900℃, so the refractories are required to have good thermal stability.

（3）良好的热稳定性。电炉炼钢从出钢到装料几分钟时间内温度急剧变化，温度由原来的1600℃左右骤然下降到900℃以下，因此要求耐火材料具有良好的热稳定性。

(4) Good slag resistance. In the process of steelmaking, slag has strong chemical erosion on refractories, so refractories should have good slag resistance.

（4）抗渣性好。在炼钢过程中，炉渣对耐火材料有强烈的化学侵蚀作用，因此耐火材料应有良好的抗渣性。

(5) High compressive strength. Refractories must have high compressive strength, because the lining is impacted by the charge during charging, by the static pressure of molten steel during smelting, by the scouring effect of steel flow during tapping, and by mechanical vibration during operation.

（5）高耐压强度。电炉炉衬在装料时受炉料冲击，冶炼时受钢液的静压，出钢时受钢流的冲刷，操作时又受机械振动，因此耐火材料必须有高的耐压强度。

(6) Low thermal conductivity. In order to reduce the heat loss and electric energy consumption of electric furnace, the thermal conductivity of refractory material is required to be poor, that is, the thermal conductivity is required to be small.

（6）低导热性。为了减少电炉的热损失，降低电能消耗，要求耐火材料的导热性要差，即导热系数要小。

4.2.2 Refractory for Roof
4.2.2 炉盖用耐火材料

In the smelting process, the working condition of furnace roof is very bad, and it is in high temperature state for a long time. It is often affected by rapid temperature change, chemical erosion of furnace gas and slag materials, mechanical vibration of lifting furnace roof and other factors. In recent years, with the expansion of electric furnace capacity and the improvement of unit power level, the use conditions of furnace roof are more and more strict, and the refractory materials for furnace roof have also changed. Refractory materials for furnace roof are as follows:

炉盖在冶炼过程中长期处于高温状态,并且经常受到温度急变的影响,受到炉气和粉末造渣材料的化学侵蚀,受到升降炉盖的机械振动作用,所以工作条件十分恶劣。近年来,随着炼钢电弧炉容量扩大与单位功率水平的提高,炉盖的使用条件变得更加苛刻,炉盖用耐火材料也随之发生变化。炉盖用耐火材料有以下几种:

(1) Silica brick. It is made of natural quartzite or quartz sand. The physical and chemical properties of silica brick for roof are shown in Table 4-1.

(1) 硅砖。由天然石英岩或石英砂加工制成,电炉炉盖用硅砖理化性能见表4-1。

Table 4-1　Physical and chemical properties of silica brick for roof

表4-1　电炉炉盖用硅砖理化性能

Index 指标	(DG) -95
SiO_2 content (not less than)/% SiO_2 质量分数(不小于)/%	95
Refractoriness (not less than)/℃ 耐火度(不小于)/℃	1710
$2kg/cm^2$ start temperature of load softening (not less than)/℃ $2kg/cm^2$ 荷重软化开始温度(不小于)/℃	1650
Apparent porosity (not less than)/% 显气孔率(不小于)/%	22
Normal temperature compressive strength (not less than)/$kgf \cdot cm^{-2}$ 常温耐压强度(不小于)/$kgf \cdot cm^{-2}$	250

Silicon brick used to be the main material of the basic EAF roof because of its high load softening temperature, high fire resistance, small quality and low price. However, the thermal stability of silica brick is poor, because its volume changes greatly when the temperature is between 180~300℃ and 300~600℃, which is caused by the crystal transformation of SiO_2. Because of the serious thermal expansion, the roof often appears convex and concave phenomenon in use, so it is rather troublesome to masonry (expansion joint must be left). For basic electric furnace, the slag resistance of silica brick is poor, and it is easy to react with lime powder. With the increase of furnace heat load, the low fire resistance of silica brick has become the main problem. The service life of silica brick furnace roof is generally no more than 50 furnaces. The dropping of the fused silica brick on the furnace wall will affect the service life of the furnace wall; the melting drop will also reduce the basicity of the slag, which makes it difficult to refine. Therefore, in addition to the acid furnace, the basic furnace roof has rarely used silica brick.

硅砖有很高的荷重软化温度、较高的耐火度，同时具有质量小、价格便宜等优点，因此曾是碱性电弧炉炉盖的主要材料。但硅砖的热稳定性差，当工作温度在 180～300℃、300～600℃ 两个温度区间时，由于 SiO_2 晶型转变，体积变化较大，所以耐急冷急热的能力很差。由于热膨胀严重，在使用中炉盖往往出现上凸下凹现象，为此砌筑时相当麻烦（必须留膨胀缝）。对于碱性电炉来讲，硅砖的抗渣性差，极易与石灰粉末作用。随着电炉热负荷的提高，硅砖的耐火度低也成为主要问题。硅砖炉盖使用寿命一般不超过 50 炉。而且硅砖的熔滴滴在炉墙上影响炉墙的使用寿命；熔滴还会降低炉渣碱度，给精炼造成困难。因而除酸性电炉外，碱性电炉炉盖已很少使用硅砖。

(2) High alumina brick. Alumina silicate refractories containing more than 46% Al_2O_3 are called high alumina bricks, whose raw material is bauxite. According to Al_2O_3 content, it is divided into three grades. Most of the factories use first-class or second-class high aluminum bricks.

(2) 高铝砖。$w(Al_2O_3)>46\%$ 的硅酸铝质耐火材料称为高铝砖，它的原料是高铝矾土矿。根据 Al_2O_3 含量分为三级。各厂大多采用一级高铝砖或二级高铝砖。

Compared with silica brick, high alumina brick has high fire resistance, good thermal stability, good slag resistance and high compressive strength. China is rich in bauxite, so it is the main refractory used for the furnace roof of alkaline electric furnace.

高铝砖与硅砖相比，具有耐火度高、热稳定性好、抗渣性好和耐压强度高等优良性能，我国矾土矿丰富，所以目前它是我国碱性电炉炉盖用的主要耐火材料。

(3) Basic brick. Alkaline brick is a new type of furnace roof brick. At present, the basic furnace roof brick used in various countries has magnesium, chrome magnesium, dolomite, magnesium aluminum and so on. They have high refractoriness and good ability of resisting iron oxide slag. Their performance is better than that of high alumina brick under severe smelting conditions, and their service life is longer. However, due to the severe deformation and high cost, it has not been widely used. At present, it is mainly used in the vulnerable parts of the furnace roof (electrode hole, smoke exhaust hole, center), and the rest parts are still made of high aluminum bricks.

(3) 碱性砖。碱性砖是较新型的炉盖用砖。现在各国使用的碱性炉盖砖就其材质而言有镁质、铬镁质、白云石质、镁铝质等。它们具有高的耐火度和良好的抗氧化铁渣的能力，在苛刻冶炼条件下的使用性能比高铝砖好，使用寿命长。但是因为变形厉害和成本高，所以还没有广泛采用。目前主要用在炉盖的易损部位（电极孔、排烟孔、中心部），其余部位仍用高铝砖。

(4) Fireclay. When building the furnace roof, the refractory mud is mixed with brine or purified water to form refractory mud. Its function is to fill the brick joint, make the masonry have good tightness, prevent gas passing through and avoid slag penetration.

(4) 耐火泥。在砌制炉盖时用耐火泥与卤水或净水调和成耐火泥浆，其作用是填充砖缝，使砌体具有良好的紧密性，防止气体通过，避免炉渣渗透。

There are several types of refractory mud, such as clay, silica, high alumina and magnesia. Their main components and physical and chemical indexes are basically the same as the corresponding refractory bricks. Refractory mud and refractory brick shall have the same physical properties and chemical composition to ensure the strength of masonry and prevent them from corroding each other under high temperature.

耐火泥有黏土质、硅质、高铝质和镁质等几类。它们的主要成分和理化指标与相应的耐火砖基本相同。使用时耐火泥应与耐火砖具有相同的物理性质和化学成分，以保证砌体的强度及防止它们在高温下互相侵蚀。

High alumina refractory mud shall be used for building furnace roof of high alumina brick, and silica refractory mud shall be used for building furnace roof of silica brick. Chrome magnesia brick is generally dry laid, because wet laid will make it powdered at high temperature.

砌制高铝砖炉盖时用高铝质耐火泥，砌制硅砖炉盖时用硅质耐火泥。铬镁砖一般采用干砌，因湿砌会使它在高温下粉化。

(5) Refractory concrete. Refractory concrete is a new type of refractory material. Compared with refractory brick, it has the advantages of simple manufacturing process, convenient use and low cost. It is suitable for mechanized production of products with complex shapes. Generally, high alumina clinker is used as aggregate in EAF. Phosphate refractory concrete with phosphoric acid or aluminum phosphate as binder. Its fire resistance can reach more than 1800℃.

(5) 耐火混凝土。这是一种新型耐火材料，它和耐火砖相比，具有制作工艺简单、使用方便、成本低等优点，并且适于机械化制作形状复杂的制品。电炉一般使用高铝质熟料为骨料。以磷酸或磷酸铝作为胶结剂的磷酸盐耐火混凝土，其耐火度可达1800℃以上。

4.2.3　Refractory for Wall and Bottom
4.2.3　炉墙、炉底用耐火材料

4.2.3.1　Magnesia
4.2.3.1　镁砂

Magnesia's main component is magnesium oxide [$w(MgO) \geqslant 85\%$], also contains a small amount of impurities (such as SiO_2 and CaO). The higher the MgO content in magnesia, the lower the impurity content and the better the quality. If the content of SiO_2 is too high, its fire resistance will be reduced. If the content of CaO is too high, it will be easy to hydrolyze and pulverize. The magnesia used in EAF is first-class metallurgical magnesia. Its composition is shown in Table 4-2.

镁砂的主要成分是氧化镁 [$w(MgO) \geqslant 85\%$]，也含有少量的杂质（如 SiO_2 和 CaO 等）。氧化镁含量越高越好，杂质含量越低越好。含 SiO_2 太高将会降低其耐火度，含 CaO 太高则易水解粉化。电炉使用的镁砂是一级冶金镁砂。其成分见表4-2。

Table 4-2　Composition of primary metallurgical magnesia
表4-2　一级冶金镁砂成分

$w(MgO)$	$w(SiO_2)$	$w(CaO)$	Ignition reduction 灼烧减量
≥87%	≤4%	≤5%	≤0.5%

4.2.3.2 Dolomite
4.2.3.2 白云石

The main components are calcium oxide and magnesium oxide, as shown in Table 4-3.
主要成分是氧化钙和氧化镁,其成分见表4-3。

Table 4-3　Main components of dolomite
表4-3　白云石主要成分

$w(MgO)$	$w(CaO)$	$w(SiO_2)$	$w(FeO+Al_2O_3)$
35%	52%~58%	0.8%	2%~3%

The refractoriness of dolomite is also above 2000℃, it can resist the erosion of basic slag, and its thermal stability is better than magnesia, but dolomite is easy to absorb water and powder, so the time from firing to using of dolomite should be shortened as much as possible.
白云石耐火度也在2000℃以上,它能抵抗碱性炉渣的侵蚀,热稳定性比镁砂好,但白云石易吸水粉化,因此尽量缩短白云石从烧成到使用的时间。

4.2.3.3 Quartz Sand
4.2.3.3 石英砂

Quartz sand is one of the main materials for lining of acid arc furnace. It is used to build furnace bottom and slope, and also used as mending material of acid EAF.
石英砂是砌筑酸性电弧炉炉衬的主要材料之一,用来砌筑炉底和炉坡,也用作酸性电弧炉的补炉材料。

Pure quartz sand is a crystal transparent body. When there are a few impurities, the color is white. The more impurities, the darker the color is. The color of quartz sand for EAF is mostly white. Its composition is shown in Table 4-4.
纯的石英砂为水晶透明体,含有少量杂质时为白色,杂质越多就越呈暗灰色,电炉用的石英砂大多是白色的。其成分见表4-4。

Table 4-4　Main components of quartz sand for EAF
表4-4　电炉用的石英砂的主要成分

$w(SiO_2)$	$w(FeO)$	$w(Al_2O_3)$
96%~97%	1%	1.3%

4.2.3.4 All Kinds of Firebrick
4.2.3.4 各种耐火砖

Clay brick is used as heat insulating brick because of its low refractoriness and low softening temperature under load. They are usually laid close to the furnace shell. Lay magnesite or other alkaline bricks next to clay bricks. Chrome-magnesite and magnesium-aluminum bricks are used less, because the raw materials used to produce them are scarce and more expensive. Silicon brick is used in acid arc furnace. Main properties of refractory for wall and bottom of EAF is shown in Table 4-5.

黏土砖因耐火度和荷重软化温度低，故用作隔热砖，砌在靠近炉壳的部位。在黏土砖的里面再砌筑镁砖或其他碱性砖。铬镁砖和镁铝砖由于原材料缺乏，价格较贵，用得较少。硅砖用于酸性电弧炉。电炉炉墙和炉底用耐火材料的主要性能见表4-5。

Table 4-5 Main properties of refractory for wall and bottom of EAF

表4-5 电炉炉墙和炉底用耐火材料的主要性能

Species 名称	Brand 牌号	Main chemical composition (not less than) 主要化学成分（不小于）	Refractoriness/℃ 耐火度/℃	Load softening temperature /℃ 荷重软化温度/℃	Show porosity/% 显气孔率/%	Compressive strength at room temperature /kgf·cm^{-2} 常温耐压强度/kgf·cm^{-2}	Bulk density /g·cm^{-3} 体积密度/g·cm^{-3}	Coefficient of thermal conductivity /kJ·(m·h·℃)$^{-1}$ 导热系数 /kJ·(m·h·℃)$^{-1}$
Clay bricks 黏土砖	NI—30	$w(Al_2O_3)=30\%$	1610	1610	28	125	2.07	2.508+ 2.299×10^{-3}t
	NI—35	$w(Al_2O_3)=35\%$	1670	1670	26	150		
	NI—40	$w(Al_2O_3)=40\%$	1730	1730	26	150		
Silica brick 硅砖	GI—94	$w(SiO_2)=94.5\%$	1710	1640	23	200	1.9	3.762+ 3.344×10^{-3}t
	GI—93	$w(SiO_2)=93.0\%$	1690	1620	25	175		
Magnesite brick 镁砖	M—87	$w(MgO)=87\%$	2000	1500	20	400	2.6	15.466− 1.714×10^{-3}t
Dolomite brick 白云石砖		$w(CaO)=40\%$ $w(MgO)=30\%$	1700~1800	1550~1610	20	1000	2.9	7.524
Magnesite chrome brick 铬镁砖	MG—12	$w(Cr_2O_3)=12\%$ $w(MgO)=48\%$	1950	1520	23	200	2.8	7.106
Magnesium aluminum brick 镁铝砖	ML—80	$w(MgO)=80\%$ $w(Al_2O_3)=5\%~10\%$	2100	1550~1580	19	350	3.0	

4.2.4 Insulation and Binder for EAF
4.2.4 电炉用绝热材料和黏结剂

4.2.4.1 Thermal Insulation Material
4.2.4.1 绝热材料

The purpose of the adiabatic material is to reduce the heat loss of the furnace linings. Common thermal insulation materials and main properties are shown in Table 4-6.

绝热材料的作用是为了减少炉衬的热损失。常用绝热材料及主要性能见表4-6。

Table 4-6　Commonly used thermal insulation materials and main properties
表4-6　常用绝热材料及主要性能

The name of the material 材料名称	Bulk density/g·cm^{-3} 体积密度/g·cm^{-3}	Permissible operating temperature/℃ 允许工作温度/℃	Coefficient of thermal conductivity/kJ·(m·h·℃)$^{-1}$ 导热系数/kJ·(m·h·℃)$^{-1}$
Asbestos board 石棉板	0.9~1.0	500	$(0.585~0.627)\times10^{-3}t_p$
Asbestos board 硅藻土	0.55	900	$0.334+0.878\times10^{-3}t_p$
Diatomite brick 硅藻土砖	0.55~0.7	900	$(0.334~0.711)+0.836\times10^{-3}t_p$
Light clay rotation 轻质黏土砖	0.4	900	$0.293+0.394\times10^{-3}t_p$

Annotation: t_p—average temperature.
注: t_p—平均温度。

The volume density of adiabatic materials is relatively small, the smaller the volume density, the lower the thermal conductivity, the better the adiabatic effect. Because the allowable working temperature of the insulation material is low, it can only be used in the place where the temperature is less than 1000℃, so it is only used in the outermost layer of the furnace wall and the bottom layer of the furnace on the EAF to avoid contact with high temperature slag and furnace gas.

绝热材料的体积密度都是比较小的，体积密度越小导热系数越低，绝热作用越好。因为绝热材料的允许工作温度较低，只能用于温度在1000℃以下的部位，所以在电炉上只用在炉墙的最外层和炉底的最下层，避免与高温炉渣和炉气接触。

At present, the basic EAF in China mainly USES asbestos board, diatomite powder and clay brick as insulation materials. Asbestos is a fibrous mineral consisting mainly of magnesium, silicon, and calcium compounds. Diatomite powder is made from diatomite and the main component is SiO_2.

我国目前碱性电炉主要使用石棉板、硅藻土粉和黏土砖作绝热材料，石棉板是用石棉和黏结剂制成的板状材料。石棉是一种纤维状的矿物，主要组分是镁、硅、钙化合物。硅藻土粉是硅藻土加工成的，主要成分是SiO_2。

4.2.4.2　Binder
4.2.4.2　黏结剂

(1) Bitumen. A black solid, tar fractionated residue. The electric furnace generally uses me-

dium temperature asphalt, which is easy to carbonize under high temperature (above 200℃). The fixed carbon left after the removal of asphalt volatiles plays the role of skeleton in the furnace lining, which can improve the fire resistance, slag resistance and thermal stability of magnesia and dolomite. Before use, the asphalt should be dehydrated so that the water content is not higher than 0.50%.

（1）沥青。沥青是焦油分馏后的残留物，为黑色固体。电炉一般采用中温沥青，它在高温下（200℃以上）易碳化。沥青挥发物去掉后留下的固定碳在炉衬中起骨架作用，可提高镁砂和白云石的耐火度、抗渣性和热稳定性。沥青使用前，需做脱水处理，使水分不高于0.50%。

（2）Tar. Tar is a byproduct of coking. It is a black, sticky liquid with a high viscosity. Tar is mainly used for knotting furnace lining and making magnesia brick (or dolomite brick) binder. Tar should also be dehydrated before use, so that the water content is not higher than 0.50%.

（2）焦油。焦油是炼焦的副产品，是黑色黏稠状液体，其黏性极大。焦油主要用作打结炉衬和制作镁砂砖（或白云石砖）的黏结剂。焦油使用前也需做脱水处理，使水分不高于0.50%。

（3）Brine. The main ingredient of brine is magnesium chloride ($MgCl_2$), usually supplied as a solid. Before use, water can be added according to the specific proportion of the requirements, after heating and melting into the requirements of the water solution. The brine is mainly used for mixing refractory mortar and as a binder for magnesia when knotting carbonless furnace lining.

（3）卤水。卤水主要成分是氯化镁（$MgCl_2$），通常以固态供应。使用前可根据要求比例加入净水，经加热熔化成符合要求的水溶液后使用。卤水主要用于拌耐火泥，以及在打结无碳炉衬时作为镁砂的黏结剂。

（4）Sodium silicate. The main component is $Na_2O \cdot nSiO_2$, generally in the liquid supply, there are also block solid sodium silicate. It contains $w(SiO_2) = 71\% \sim 76\%$ and $w(Na_2O) = 8\% \sim 14\%$. Sodium silicate is mainly used for mixing refractory mortar and in the formation of acid furnace lining to do quartz sand binder.

（4）水玻璃：水玻璃（$Na_2O \cdot nSiO_2$）又称硅酸钠或泡花碱，一般以液态供应，也有块状固体水玻璃。它含有$w(SiO_2) = 71\% \sim 76\%$，$w(Na_2O) = 8\% \sim 14\%$。水玻璃主要用于拌耐火泥，以及在打结酸性炉衬时做石英砂的黏结剂。

Exercises

(1) What are the requirements of electric arc furnaces for refractory materials?

(2) What are the types of refractory materials used in various parts of the electric arc furnace?

思考题

（1）电弧炉对耐火材料的要求有哪些？

（2）电弧炉各部位使用的耐火材料种类是什么？

Task 4.3　Masonry of Furnace Lining
任务4.3　炉衬的砌筑

Mission objectives
任务目标

（1）Understand the structure of the lining of each part of the electric arc furnace.
（1）了解电弧炉各部分炉衬的结构。
（2）Master the building method of electric arc furnace linings.
（2）掌握电弧炉炉衬的砌筑方法。

4.3.1　Masonry of Furnace Bottom
4.3.1　炉底的砌筑

The structure of electric furnace bottom is divided into three layers: insulation layer, protective layer (bricklaying layer) and working layer.

电炉炉底的结构分为绝热层、保护层（砌砖层）和工作层3层。

4.3.1.1　Thermal Barrier
4.3.1.1　绝热层

The insulation layer is the lowest layer at the bottom of the furnace. Its function is to reduce the heat loss of the electric furnace and ensure the small temperature difference between the molten steel above and below the molten pool. The usual laying method is: on the furnace shell, first lay a layer of asbestos plate (about 10mm), and then lay diatomite powder (≤20mm), diatomite powder above a layer of flat insulation brick (diatomite brick or clay brick: 65mm), brick joints should not be greater than 2mm, gaps with diatomite powder or clay brick powder filled.

绝热层是炉底的最下层，它的作用是减少电炉的热损失，并保证熔池上下钢液的温差小。通常的砌法是：在炉壳上，先铺一层石棉板（约10mm），再铺硅藻土粉（≤20mm），硅藻土粉上面平砌一层绝热砖（硅藻土砖或黏土砖：65mm），砖缝应不大于2mm，缝隙用硅藻土粉或黏土砖粉填充。

4.3.1.2　Protective Layer
4.3.1.2　保护层

Above the insulation is a protective layer. Its role is to ensure the pool of robustness, prevent steel leakage. The protective layer is made of fine alkaline brick, usually made of magnesia brick.

绝热层上面是保护层。其作用是保证熔池的坚固性，防止漏钢。保护层采用优良的碱性砖，一般用镁砖砌筑。

There are two types of magnesia bricks, including shaped bricks and standard bricks. When using special-shaped bricks, the brick joints are small and the quality is good. Standard bricks, however, are not as good, but they also ensure the quality of operation when finished with a brick grinder.

所用的镁砖有异型和标型两种。采用异型砖时，砌制砖缝小，质量好，而标准砖差些，但使用磨砖机修整后，也能保证操作质量。

There are three ways of laying bricks: flat, side and vertical. Flat laying is easy to operate, but the brick surface is easy to arch when heated. Lateral and vertical masonry is more difficult, but not easy to arch after the heat. Electric ovens are generally flat.

砌砖方法有3种，那平砌、侧砌和立砌（竖砌）。平砌操作简单，但砖面较大时受热后容易拱起；侧砌和立砌比较困难，但受热后不容易拱起。电炉一般采用平砌。

When laying bricks, first lay 2~4 layers of 65mm thick magnesia bricks on the insulating brick layer. Then lay 4~6 layers of 65mm thick magnesia brick along all sides, lay them into a stepped or cylindrical shape. Layer to layer to cross the system, each other into 60° or 45° angle, not parallel or at right angles; all layers of brickwork shall be flat and tight, with small joints (≤2mm), and the joints shall be filled with magnesia powder with a particle size of no more than 0.5mm.

砌制时，首先在绝热砖层上平砌2~4层65mm厚的镁砖。然后再沿四周平砌4~6层65mm厚的镁砖，砌成台阶形或圆柱形。层与层之间要交叉砌制，互相成60°或45°角，不得平行或成直角；各层砌砖都要平整、紧固，砌缝要小（≤2mm），砖缝用粒度不大于0.5mm的镁砂粉填充，并以手锤敲击砖面，使其填充密实。

After insulation and protective layer are laid, it should be kept dry.

绝热层和保护层砌成后，应保持干燥。

4.3.1.3　Work Layer

4.3.1.3　工作层

Above the protective layer is the working layer. The working layer contains molten steel and slag, which is directly in contact with steel slag. The thermal load is high, the chemical erosion is serious, the mechanical erosion is strong, and the damage is extremely easy. Therefore, its quality must be guaranteed.

保护层上面是工作层。工作层是容纳钢液和炉渣的部位，直接与钢渣接触，热负荷高，化学侵蚀严重，机械冲刷作用强烈，极易损坏，因此必须保证它的质量。

The refractory used in the working layer is magnesia, tar and asphalt used as binder. When making carbon-free furnace lining, use brine or sodium silicate as binder.

工作层使用的耐火材料为镁砂，黏结剂用焦油、沥青；制作无碳炉衬时，用卤水或水玻璃作黏结剂。

There are three forming methods of working layer: knot, vibration and masonry.

工作层成形方法分打结、振动和砌筑 3 种。

(1) Knot forming. The grain size of magnesia for knotting should be properly proportioned. Generally, magnesia is used in combination with grain size of 3~8mm 70% and 1~3mm 30%. Large magnesia plays a backbone role to ensure the strength of the working layer, while small magnesia fills the gap to ensure its tightness. Magnesia should be baked at temperatures above 200℃ to remove moisture before use. Magnesia, tar and bitumen should be evenly mixed at the right temperature. The knotting temperature is generally 100~130℃, and the pressure of the wind hammer used for knotting is 0.6~0.7MPa. The knot should be layered, the first layer with 30~40mm is appropriate, after each layer should be less than 20mm, 20t furnace about 300mm.

(1) 打结成形。打结用镁砂的粒度应有恰当的配比。通常镁砂按粒度 3~8mm 占 70%、1~3mm 占 30% 组合使用，效果较好。大粒镁砂起骨干作用，保证工作层的强度，小粒镁砂填充空隙，保证它的紧密性。镁砂使用前应经 200℃ 以上温度烘烤，以除去水分。镁砂和焦油、沥青应在合适的温度下均匀混合。打结温度一般为 100~130℃，打结使用的风锤压力为 0.6~0.7MPa。打结要分层进行，第一层以 30~40mm 为宜，以后各层应少于 20mm，20t 炉子约 300mm。

The biggest characteristic of knotting furnace bottom is integrity and no gap. The bottom of a knotted furnace shall be hemispherical in shape, the size shall meet the requirements, and the tap hole shall have enough slope.

打结炉底的最大特点是整体性，没有缝隙。一个打结好的炉底应为半球形，尺寸合乎要求，出钢口处具有足够的坡度。

(2) Vibration forming. Manual knotting is characterized by low efficiency, poor working conditions, high labor intensity and no guarantee of knotting quality. The principle of vibration forming is that under the action of vibration of higher frequency and small amplitude, small particles and fine powder will like a liquid into the gap between the large particles, thus increasing the density of the material.

(2) 振动成形。人工打结成形，工作效率低，劳动条件差，劳动强度大，打结质量得不到保证。振动成形的原理在于：材料在较高的频率和小振幅的振动作用下，小粒和细粉会像液体一样钻进大粒间的缝隙中，从而提高了材料的密度。

The raw material preparation for vibration forming is generally similar to manual knotting, but the ratio should be different.

振动成形的原材料准备大体与人工打结相似，但配比应有所区别。

(3) Masonry forming. In addition to the use of vibration forming, some factories have adopted the method of structuring the working layer of the furnace bottom with small bricks. After the raw materials are mixed uniformly in proportion, the bricks are prefabricated into small bricks, and then built according to the size of the furnace.

(3) 砌筑成形。除了采用振动成形外，有些厂采用了机制小砖砌筑炉底工作层的方法。各种原材料按比例均匀混合后，用压砖机预制成小砖，然后根据炉体尺寸要求进行砌筑。

Asphalt magnesia bricks for the bottom of the furnace, and brine magnesia bricks for the carbon-free lining. The size of various bricks should be consistent with the size of the furnace shell to ensure that the bricks are tightly closed and easy to build. The shape of the brick must be complete without corners or edges. And it must be properly kept to prevent moisture. When laying bricks, the brick joints should be staggered and cannot overlap, and the brick joints should not be larger than 2mm. The gap should be plugged with packing.

炉底用沥青镁砂砖，无碳炉衬用卤水镁砂砖。各种砖的尺寸应与炉壳尺寸相吻合，以保证砌砖时砖缝紧密，砌筑方便。砖的外形要完整，不能有缺角缺棱。而且要妥善保管，防止受潮。砌砖时，砖缝要错开，不能重叠，砖缝不大于2mm，缝隙用填料塞紧。

4.3.2 Masonry of Furnace Walls
4.3.2 炉墙的砌筑

The structure of furnace wall can be divided into two parts: insulation layer and working layer. The insulation layer is constructed of asbestos sheet and clay brick. Usually a layer of 10 ~ 15mm asbestos plate is used in the place of close to the furnace shell steel plate, and then a layer of 65mm clay brick is built to form the insulation layer. In order to enlarge the load of electric furnace and reduce the collapse of upper furnace wall, some factories cancel the insulation layer of furnace wall and replace it with a thin layer of asphalt magnesia pressed brick. This will reduce the total thickness of the furnace wall but increase the working layer, and make the upper furnace wall has the opportunity to bond with the furnace shell steel plate.

炉墙的结构一般可分为保温层和工作层两部分。保温层用石棉板和黏土砖砌筑。通常在紧贴炉壳钢板处用一层10~15mm的石棉板，再砌筑一层65mm的黏土砖，构成保温层。有些厂为了扩大电炉装入量和减少上部炉墙的倒塌，取消炉墙保温层，用一层薄型的沥青镁砂压制砖代替，使炉墙总厚度减薄但工作层增加，并且上部炉墙有机会与炉壳钢板烧结牢。

There are three kinds of masonry methods for the working floor:

工作层的砌筑方法可分3种：

(1) Made of alkaline bricks or small bricks which are made by machining. With magnesia brick, magnesia aluminum brick and chrome-magnesia brick masonry furnace wall cost the highest, general domestic steel mills are rarely used. But these furnace walls have a high degree of fire resistance, and will not make the steel fluid carbon, so some steel mills smelting low carbon or ultra-low carbon steel, still in use.

(1) 用碱性砖或机制小砖砌筑。用镁砖、镁铝砖和铬镁砖砌筑的炉墙成本最高，一般国内钢厂很少采用。但这些炉墙有较高的耐火度，且不会使钢液增碳，所以某些钢厂在冶炼低碳或超低碳等钢种时，仍在使用。

The small mechanical bricks used for masonry furnace wall are asphalt magnesia brick, dolomite brick, brine magnesia brick. Carbonless furnace lining with brine magnesia brick, slag furnace with stone grinding silica brick masonry. The density of small brick is higher than that of knotted furnace wall, which has higher service life and better working conditions.

砌筑炉墙所用的机制小砖有沥青镁砂砖、白云石砖、卤水镁砂砖。无碳炉衬用卤水镁砂砖，化渣炉用石磨硅砖砌筑。机制小砖的密度高于打结炉墙，有较高的使用寿命，劳动条件也较好。

(2) Bulk magnesite brick assembly. When adopting this method, divide the furnace wall into 3 large blocks and 1 small block (on the furnace door frame). The methods of making large magnesia are manual knotting and vibration forming. Manual knotting is done in a prefabricated steel mold, which is small on top and large on bottom, with a gradient. The knotting material and knotting process are the same as the bottom of the furnace. When the mold is placed on the vibration table, the mixture can be added in batches or continuously, and then the pressure block is placed to vibrate. When the furnace wall vibrates and forms, the vibration from the beginning to the stripping shall not be less than 8h, and the volume density shall be greater than 2.86kg/cm^3.

(2) 大块镁砂砖装配。采用这种方法时，将炉墙分成3大块和1小块（炉门框上）。大块镁砂制作的方法：一种是人工打结；另一种是振动成形。人工打结是在预制的钢模中进行，模子上小下大，有一个坡度。打结材料和打结过程与炉底相同。振动成形时模子放在振动台上，混合料可分批或连续加入，然后放上压块进行振动，炉墙从振好到脱模不应少于8h，体积密度应大于2.86kg/cm^3。

(3) Integral knotting and integral vibration forming. The material, ratio, technical operation and quality requirements of knotting and vibration forming are the same as those of furnace bottom. In order to save magnesia, some or all of the waste magnesia or dolomite can be used in the upper part of the furnace wall. Furnace wall can also be all dolomite knot.

(3) 整体打结和整体振动成形。打结和振动成形的材料、配比、技术操作和质量要求等与炉底相同。为了节省镁砂，在炉墙的上部可采用一部分或全部废镁砂或白云石，在不同的高度上可以采用不同质地的材料。炉墙也可全部采用白云石打结。

The thickness of the furnace wall shall be determined according to the actual situation, and the external temperature of the furnace shell shall not be greater than 150℃. The thickness of slag line at the bottom of furnace wall is similar to that at the center of furnace bottom.

炉墙的厚度按实际情况而定，应能保证炉壳外部温度不大于150℃为好。炉墙下部渣线处的厚度与炉底中心处的厚度相近。

4.3.3 Masonry of Roof
4.3.3 炉盖的砌筑

At present, the basic EAF roof mostly uses a high aluminum brick, which is divided into two types: special-shaped brick and standard brick. Adopt standard brick, cut brick quantity is big, time-consuming, masonry quality is not high. And use hetero-shaped brick, the situation is exactly the opposite, cut brick quantity is less, brick seam is small, masonry quality is good, so the cost of work is less. Therefore, China's steel mills are using shaped bricks.

目前，碱性电弧炉炉盖大多采用一级高铝砖，分为异型砖和标准砖两种。采用标准砖，砍砖量大，费工费时，砌筑质量不高。而采用异型砖，情况则恰恰相反，砍砖量较少，砖缝小，砌筑质量好，所以费工时也较少。因此，我国各钢厂均采用异型砖。

There are many masonry methods of roof, the most common is the 'herringbone method'.
炉盖的砌筑方法很多，最普遍的是"人字形砌法"。

Before masonry, we should check whether the roof ring is deformed, and do water pressure test. If there is deformation and leakage, it should be used after correction and repair. In order to ensure a certain degree of arch after the roof is built, we should put the roof ring on a concrete or steel mold before masonry, place the open hearth roof ring, and start the masonry after centering. In the herringbone masonry method, the '+' word line is usually drawn from the center of the furnace cover. One of the cross lines must pass through the center of No. 2 electrode hole. First lay the cross brick, then from the bottom up to the middle to the electrode hole, then start laying the three electrode holes, and finally lay the electrode around and in the center. There are also first laying electrode hole. The clay used in masonry is a mixture of high alumina brick powder and brine. The dimensions of each part of the roof shall conform to the requirements of the drawings. The brick joints shall be no more than 2mm, and the difference between the height and height of the brick shall be no more than 5mm. After the completion of masonry, the roof shall be dried.

炉盖砌筑前，应检查炉盖圈是否变形，并做水压试验，如有变形和漏水现象，则应在矫正和修补后才能使用。为了保证炉盖砌好后有一定的拱度，砌筑前应把炉盖圈放在混凝土或钢制的模子上，炉盖圈要摆平，找正中心后再开始砌筑。人字形砌筑法砌筑时，一般首先以炉盖中心为基点拉"十"字线，此"十"字线中一根必须通过2号电极孔的圆心。先砌好十字线砖，然后由此从下而上向中部铺砌至电极孔部位，再开始砌3个电极孔，最后砌电极周围及中心部位，也有先砌电极孔的。砌筑时用的泥料是高铝砖粉和卤水的混合物。炉盖各部分的尺寸应符合图纸的要求，砖缝不大于2mm，砖与砖高低凸凹差不大于5mm，砌筑完毕后应进行干燥。

The roof can also be made of fire-resistant concrete. The overall knotted roof can reach the level of the first and second grade high-alumina brick roof, and opens a new way to solve the materials used in the roof.

炉盖还可以用耐火混凝土制作。这种整体打结的炉盖可以达到一、二级高铝砖炉盖的水平，并为解决炉盖材料开辟了新途径。

Exercises

(1) What is the structure of the furnace bottom? What are the masonry methods for the working floor of the furnace bottom?

(2) How is the working layer of the furnace wall built?

(3) How is the roof built?

思考题

(1) 炉底的结构是怎样的，炉底工作层有哪些砌筑方法？

(2) 炉墙的工作层是如何砌筑的？

(3) 炉盖是如何砌筑的？

Task 4.4　Maintenance of Furnace Lining
任务4.4　炉衬的维护

Mission objectives
任务目标

（1）Understand the cause of furnace lining damage.
（1）理解炉衬损坏的原因。
（2）Master the method of furnace lining maintenance.
（2）掌握炉衬维护的方法。

4.4.1　Causes of Lining Damage
4.4.1　炉衬损坏的原因

The erosion of furnace lining by slag depends on its composition and fluidity. The higher the content of SiO_2 in slag, the more serious the erosion of alkaline furnace lining. In addition to SiO_2, Al_2O_3, Fe_2O_3 and CaF_2 in slag also corrode furnace lining. Slag fluidity also affects lining. The thin slag has a serious scouring effect on the lining, and the ability of reflecting arc is strong, which will increase the heat load of the lining. The thick slag makes it difficult to heat the molten pool, and the physicochemical reaction conditions become worse, which prolongs the high-temperature action time of furnace lining (i. e. smelting time). These will accelerate the lining damage.

炉渣对炉衬的侵蚀作用，取决于它的组成和流动性。渣中 SiO_2 含量越高，碱性炉衬被侵蚀越严重。除 SiO_2 外，渣中 Al_2O_3、Fe_2O_3、CaF_2 等，同样侵蚀炉衬。炉渣流动性对炉衬也有影响，渣稀对炉衬冲刷作用严重，并且反射电弧的能力强，会增加炉衬的热负荷；渣稠使熔池加热困难，物化反应条件变差，延长炉衬高温作用时间（即冶炼时间）。这些都会加速炉衬的损坏。

In many cases, the damage of furnace lining is caused by the impact of mechanical forces, such as unreasonable distribution of burden, the bottom of the furnace without the protection of small scrap steel, charging when the tank is too far away from the furnace, the bottom of the furnace may be large scrap steel into pits. The mechanical vibration caused by roof lifting and furnace body opening is also the cause of damage to furnace lining. In addition, the furnace charge ratio is improper, smelting operation is not correct, such as oxygen blowing when the high temperature oxygen flow directly touches the furnace lining, will damage the furnace lining.

在许多情况下，炉衬的损坏是由于机械力的冲击而引起的，如布料不合理、炉底没有小废钢保护、装料时料罐太高，炉底就可能被大块废钢砸成凹坑。炉盖升降，炉体开出引起的机械振动也是损坏炉衬的原因。另外，炉料配比不当，冶炼操作不正确，如吹氧时高温的氧气流股直接触及炉衬等，都会损坏炉衬。

4.4.2 Furnace Bottom and Slope Maintenance
4.4.2 炉底和炉坡的维护

The molten steel and slag should be poured out every time the furnace is tapped. When there are residual steel and slag at the bottom of the furnace, it should be pulled out as soon as possible to avoid reducing the refractoriness or making the bottom of the furnace rise. When there are pits and rises at the bottom, they should be dealt with in time. When charging, the time should be shortened as much as possible, to prevent the furnace from severe impact on the furnace. Before charging, sufficient lime should be added to the bottom of the furnace uniformly; a person in charge should be responsible for charging. The material tank should not be too high from the bottom of the furnace to prevent the material block from damaging the bottom and slope of the furnace.

每炉出钢应将钢水和炉渣全部倒净，炉底有残钢和残渣时应尽快扒出，以免降低耐火度或使炉底上涨，必须始终保持炉底、炉坡的正确形状，当炉底出现凹坑和上涨时，应及时处理。装料时应尽量缩短时间，要防止炉料对炉子的严重冲击。装料前炉底应均匀加入足够量的石灰；装料时应由专人负责，料罐不能距炉底太高，以防止料块撞坏炉底、炉坡。

During the melting period, it is necessary to prevent the furnace material from 'bridging', and the 'bridging' must be processed in time to prevent the molten steel from overheating and eroding the bottom of the furnace. When oxygen blowing is used for fluxing, the high-temperature oxygen stream is not allowed to touch the furnace bottom and furnace slope; the solid charge on the furnace slope should be pulled into the melting pool as much as possible. The solid charge that cannot be pulled into the molten pool is cut with oxygen, but the solid charge in the molten pool cannot be interrogated with an oxygen blowing tube.

熔化期应防止炉料"搭桥"，"搭桥"一发生就必须及时处理，防止下层钢液过热熔蚀炉底。吹氧助熔时，不准高温的氧气流股触及炉底、炉坡；炉坡上的固体炉料，应尽量将其拉入熔池内，万不得已时，才用氧气切割，不得用吹氧管探询熔池中的固体炉料。

During oxygen decarburization, the insertion depth of the oxygen blowing pipe should be appropriate. It must swing back and forth, left and right, and must not touch the bottom and slope of the furnace to prevent local overheating and damage. When adding ore, it should be evenly distributed to prevent the molten pool from boiling too violently and damaging the furnace bottom and furnace slope.

氧化期吹氧脱碳时，吹氧管插入深度应合适，必须前后左右来回摆动，不准触及炉底和炉坡，以防局部过热而损坏。加入矿石时，应均匀散布，以免熔池过于激烈沸腾，损坏炉底、炉坡。

During the smelting process, the power supply should be strictly in accordance with the power curve, and the temperature of each period should be mastered; the alkalinity and fluidity of the slag should be adjusted at all times to prevent the alkalinity from being too low and the slag from being too thin and eroding the lining. When stirring the molten pool, care should be taken that the iron rake does not touch the bottom of the furnace. Especially when smelting tungsten-containing steel, be careful when testing the solubility of tungsten wire in the furnace bottom.

冶炼过程中，应严格按照电力曲线供电，掌握好各期温度；应时刻注意调整炉渣的碱度和流动性，防止碱度过低和炉渣过稀，侵蚀炉衬。搅拌熔池时，应注意铁耙不要碰炉底，特别是炼含钨钢时，向炉底试探钨丝溶解程度时要小心。

In short, all abnormal operations should be avoided and the smelting time should be shortened as much as possible.

总之，应避免一切不正常操作，尽量缩短冶炼时间。

4.4.3 Furnace Wall Maintenance
4.4.3 炉墙的维护

The life of the furnace wall is mainly determined by local damage. Because the lower part of the furnace wall is immersed in high-temperature slag for a long time, the chemical corrosion and physical penetration of the slag on the furnace wall are very serious, and the fluidity erosion of steel and slag is also relatively strong. The combined effect of the two creates a deep groove in the furnace wall, commonly referred to as the 'slag line'. The erosion of the slag line is closely related to the slag system. The temperature distribution of each part of the furnace wall is uneven. In the area near the three-phase electrode, due to the closest to the arc, the temperature is the highest and the damage is the fastest. It is usually called a 'hot spot'. Especially in the furnace wall near the No. 2 electrode, the erosion is extremely serious, especially the slag line part, which is often the weak link of the furnace wall life. There are two main measures to reduce the loss in hot spots. One is to improve the heat source. For example, the power of the three-phase arc can be balanced; select the appropriate electrode position and electrode center circle (the electrode should be perpendicular to the bottom of the furnace, or slightly inclined towards the center of the furnace); the reduction period should be operated with a low voltage short arc. The second is to improve the furnace lining. For example, high-quality refractories with high refractoriness, good slag resistance, and low porosity can be selected to build hot spots; use water-cooled slag furnace walls, etc.

炉墙的寿命主要决定于局部损坏。炉墙下部由于长时间浸泡在高温炉渣中，炉渣对炉墙的化学侵蚀和物理渗透都非常严重，加上钢、渣的流动性冲刷，造成一道深深的凹槽，通常称之为"渣线"。渣线的侵蚀与造渣制度有密切关系。炉墙各部位温度的分布是不均匀的，在靠近三相电极的区域，由于离电弧最近，温度最高，损坏最快，通常称之为"热点"。尤其是2号电极附近的炉墙，特别是渣线部位，蚀损极严重，常是炉墙寿命的薄弱环节。降低热点区域损耗的措施主要有两个方面：一是改进热源。可使三相电弧功率平衡；选择合适的电极位置及电极圆心圆。电极应垂直炉底，或向炉子中心稍微倾斜一点；

还原期尽量用低电压短电弧操作。二是改进炉衬。可选择耐火度高、抗渣性好、气孔率低的优质耐火材料砌筑热点区域；使用水冷挂渣炉壁等。

Correct charging, oven, reasonable power supply and shortening smelting time are all important links to improve the life of the furnace wall. In addition, shorten the charging time of furnace replenishment and prevent rapid cooling of the furnace wall; proper oxygen blowing operation should be paid attention to.

正确装料、烘炉、合理供电和缩短冶炼时间等都是提高炉墙寿命的重要环节。此外，缩短补炉装料时间，防止炉墙急冷；正确吹氧操作等都应加以重视。

4.4.4 Roof Maintenance
4.4.4 炉盖的维护

According to the observation, the damage of the high aluminum brick roof is the result of the combined action of chemical erosion and spalling, mainly chemical erosion, and spalling usually occurs after the brick body is eroded. The damage degree of each part of the entire roof is also different. The middle part of the roof (near the electrode hole) is the most severely damaged, followed by the brick (arch brick). This is because the high temperature effect and the chemical erosion of smoke and dust are very serious near the electrode hole, and this part uses more cut bricks in the masonry, and the brick joints are concentrated, uneven, and most eroded; Seriously affected by mechanical vibration and rapid cooling. This local damage determines the life of the roof, and sometimes makes the roof collapse.

根据观察，高铝砖炉盖的损坏是化学侵蚀与剥落共同作用的结果，以化学侵蚀作用为主，而剥落往往是在砖体被侵蚀后才发生的。整个炉盖各部位损坏的程度也是不同的，炉盖中部（电极孔附近）损坏最为严重，其次是座砖（拱角砖）附近。这是因为电极孔附近受高温作用和烟尘的化学侵蚀极为严重，并且这部分在砌筑时砍砖多、砖缝集中、不平整，最容易侵蚀；而座砖附近受机械振动和急冷作用严重。这种局部的损坏就决定了炉盖的寿命，并有时会产生炉盖塌落现象。

After tapping, the furnace replenishment and charging time should be shortened as much as possible to reduce heat loss, thereby reducing the quenching effect on the roof and preventing flaking. When loading, the furnace body should not be prematurely opened or the roof rotated. When the charge is higher than the furnace body, it must be flattened with a heavy object to avoid the roof being damaged by the charge. When lowering the roof of the furnace, it should be operated carefully, and it should not be forced too hard to prevent the loose or collapse of the roof. When lifting the electrode, attention must be paid to avoid the roof being damaged by the electrode or being damaged by the electrode sealing ring.

出钢后，应尽量缩短补炉、装料时间，减少热损失，减少对炉盖急冷作用，防止剥落。装料时，炉身开出或炉盖旋转不能过早。炉料高出炉身时，必须用重物压平，避免炉盖被炉料撞坏。降落炉盖时，应细心操作，不能用力过猛，防止震松或震垮炉盖。升降电极时，必须注意力集中，避免炉盖被电极憋坏或因电极密封圈被电极带起碰损。

To avoid big boiling during the oxidation period, when there is a sign of big boiling, the power must be cut off in time and the ore addition must be stopped. The life of the roof is closely related to the smelting time. Long smelting time, especially the long reduction period, will significantly reduce the life of roof. Smelting low-carbon steel and high-alloy steel requires special attention to shorten the smelting time, because the roof works at a very high temperature. The fluidity of the slag should be suitable. It is best to make foamed slag to reduce the reflection of the arc. Powdered slag making materials should be used less and not to reduce the chemical attack on the roof. The temperature should be controlled normally to avoid heating up after the reduction period and excessive temperature of tapping. When loading, some small material must be installed in the upper part in order to penetrate the well, so that the arc is quickly buried in the material, so as to prevent the arc from locally overheating and damage the roof. When the power is turned on, the intermediate voltage should be used, and then the high voltage power supply should be changed. It is forbidden to use high voltage during the reduction period.

氧化期要避免大沸腾，在出现大沸腾征兆时，必须及时停电，停止加矿。炉盖寿命的长短与冶炼时间有密切关系。冶炼时间长，特别是还原期时间长，炉盖寿命显著下降，冶炼低碳钢和高合金钢后，炉盖在非常高的温度下工作，尤其要注意缩短冶炼时间。炉渣的流动性应该合适，最好造泡沫渣，以减少对电弧的反射。粉状造渣材料应少用和不用，以减少对炉盖的化学侵蚀。温度应控制正常，避免还原期后升温和出钢温度过高。装料时上部必须装些小料以便穿井，使电弧很快地埋入料中，以免弧光使炉盖局部过热而损坏。开使通电时，应采用中级电压，然后再改高电压供电。在还原期禁止使用高电压。

Ash deposits on the roof hinder heat dissipation. The ash on the roof must be blown out before each furnace is tapped, which is also beneficial to ensure the quality of the steel.

炉盖积灰妨碍散热，每炉出钢前必须吹净炉盖灰，这对保证钢的质量也是有好处的。

4.4.5 Water-cooled Furnace Lining
4.4.5 水冷炉衬

4.4.5.1 Water-cooled Slag Furnace Wall
4.4.5.1 水冷挂渣炉壁

As the unit power level of EAF increases, the heat load received by the lining increases sharply, and the imbalance of the temperature distribution in the furnace increases, greatly reducing the service life of the lining refractory. The water-cooled furnace wall solves the melting loss of refractory materials caused by the strong radiation of the arc on the furnace wall during high-power operation. The water-cooled furnace wall technology has become an important part of UHP technology. The water-cooled furnace wall is a long-life furnace wall that UHP can replace the refractory lining. The average water-cooled furnace wall area of modern electric arc furnaces has reached more than 70%, as shown in Figure 4-1.

由于电弧炉单位功率水平的提高，炉衬接受的热负荷急剧增大，炉内温度分布的不平衡加剧，大大降低了炉衬耐材的使用寿命。水冷炉壁解决了高功率操作时电弧对炉壁强烈辐射引起的耐火材料的熔损，水冷炉壁技术已成为 UHP 技术的重要组成部分，水冷炉壁是 UHP 可替代耐火材料炉衬的长寿炉壁，现代电弧炉的平均水冷炉壁面积已达到 70% 以上，如图 4-1 所示。

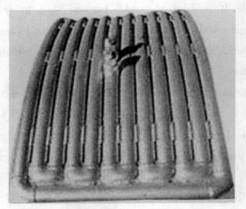

Figure 4-1　Water-cooled slag furnace wall
图 4-1　水冷挂渣炉壁

　　There are two types of water-cooled slag furnace walls, casting and welding. Castings can be divided into cast steel and embedded steel pipe cast iron. The welding slag block is usually welded into a box shape with a low carbon steel plate with a thickness of 15～20mm. There is a deflector in the cavity to ensure the cooling effect of the water flow. Whether it is a casting or welded slag hanging block, there are hanging ribs on its heated surface to fill refractory materials and improve the slag hanging capacity.

　　水冷挂渣炉壁有铸造和焊接两类。铸造件又可分铸钢和内埋钢管铸铁件两种形式。焊接挂渣块常采用 15～20mm 厚的低碳钢板焊接成箱形，空腔内有导流板以保证水流冷却作用。无论是铸造件或焊接件挂渣块，在其受热面上都设有挂扎筋，以填塞耐火材料和提高挂渣能力。

　　The principle and process of the water-cooled slag furnace wall is: at the beginning of use, the surface temperature of the slag block is much lower than the temperature in the furnace, and the slag, soot and the surface of the water-cooled block will solidify quickly, and the surface of the water-cooled block will gradually hang a protective layer composed of slag and dust. When the thickness of the slag layer continues to increase until its surface temperature gradually rises to the melting temperature of the slag layer, the thickness of the slag layer remains relatively stable. If the thermal load of the slag wall is further increased, the slag layer will automatically melt and thin until all peeled off. Due to the water cooling effect of the slag block, the surface temperature of the slag layer will rapidly decrease, and the slag and smoke will re-hang on the slag block the surface solidifies and thickens. In short, the slag layer on the heating surface of the water-cooled block is controlled by its own heat balance, and it spontaneously maintains a certain balance thickness, so that the life of the water-cooled furnace wall is long.

水冷挂渣炉壁的原理和过程是：使用开始时，挂渣块表面温度远低于炉内温度，炉渣、烟尘与水冷块表面接触就会迅速凝固，结果就会使水冷块表面逐渐挂起一层由炉渣和烟尘组成的保护层。当挂渣层的厚度不断增长，直至其表面温度逐渐升高到挂渣的熔化温度时，挂渣层的厚度保持相对稳定态。如果挂渣壁的热负荷进一步增加，挂渣层会自动熔化、减薄直至全部剥落，由于挂渣块的水冷作用，致使挂渣层表面温度迅速降低，炉渣和烟尘又会重新在挂渣块表面凝固增厚。总之，水冷块受热面的挂渣层受其自身的热平衡控制，自发地保持一定的平衡厚度，使水冷炉壁寿命长久。

4.4.5.2 Water-cooled Roof
4.4.5.2 水冷炉盖

The structure of EAF water-cooled roof is mainly tubular, as shown in Figure 4-2. According to the layout of water cooling pipes, the structure is divided into tubular ring, tubular ferrule and outer ring ferrule combination, tube ring and refractory combination.

电弧炉水冷炉盖的结构主要是管状的，如图4-2所示。根据水冷却管的布置，将其结构分为管式环状、管式套圈和外环套圈组合式、管式环状与耐火材料组合式等。

Figure 4-2　Water-cooled roof
图4-2　水冷炉盖

Water-cooled roof opens several holes as needed, including 3 electrode holes, auxiliary material holes, gas discharge holes, etc. Three-phase AC arc furnace requires 3 electrode holes, DC arc furnace can have 1 or 3 holes. Since the electrode itself is heated, the electrode hole should be composed of a metal ring composed of a water cooling tube with good thermal conductivity.

水冷炉盖根据需要开设数个孔，包括3个电极孔、装辅助料孔、气体排放孔等。三相交流电弧炉需3个电极孔，直流电弧炉可以有1个或3个孔。由于电极自身被加热，电极孔应由具有良好导热性的水冷却管组成的金属环构成。

(1) Combined water-cooled roof of tube ferrule and outer ring ferrule. The components of the water-cooled roof are: a tubular annular outer ring with a center, and a set of tubular support rods extending from the outer ring to the center; a tubular annular inner ring fixed on the support rod and forming several channels; the inner ring elbows formed by the concave tube parts form channels, and the inner convex tube elbow members are installed on the support rods between the channels, and form three tubular through holes (electrode holes) with the respective concave tube parts; concentric with the outer tube, It is located in the inner ferrule and consists of a tube connected to the inner ferrule to form a central ferrule, which is connected to the circulation system in the outer ferrule, and a set of cooling tube ferrules, which communicate with the support rod.

（1）管式套圈和外环套圈组合式水冷炉盖。水冷炉盖的部件有：管式环形外套圈，具有1个中心，1组管式支杆从外套圈向中心伸展；管式环形内套圈，固定在支杆上且形成数个通道；由各自凹管件构成的内套圈弯头形成通道，在通道间各自内凸管弯头构件安装在支杆上，且与各自凹管件共同形成3个管状通孔（电极孔）；与外套管同心，并位于内套圈内，由与内套圈相连的管组成中心套圈，它与外圈中的循环系统相连，还有一组冷却管套圈，它和支杆相互连通。

(2) Tubular ring and refractory combined water-cooled roof. The roof water cooling pipe is arranged as a ring water cooling, and the central part of the roof uses refractory materials, in which refractory electrode holes are provided. The roof structure is simple, and its shortcoming is short life.

（2）管式环状与耐火材料组合式水冷炉盖。炉盖水冷管布置为环状水冷，炉盖中心部分使用耐火材料，其中设置耐火材料电极孔。该炉盖结构简单，其缺点是寿命短。

Exercises

(1) What are the causes of furnace lining damage?
(2) How to increase the service life of furnace bottom and furnace slope?
(3) How to increase the service life of the furnace wall?
(4) How to increase the service life of roof?
(5) How does the water-cooled slag hanging furnace wall increase the service life of the furnace lining?

思考题

（1）炉衬损坏的原因有哪些？
（2）如何提高炉底和炉坡的使用寿命？
（3）如何提高炉墙的使用寿命？
（4）如何提高炉盖的使用寿命？
（5）水冷挂渣炉壁是如何提高炉衬使用寿命的？

References
参 考 文 献

[1] 高泽平. 炼钢工艺学 [M]. 北京：冶金工业出版社，2008.

[2] 沈才芳，等. 电弧炉炼钢工艺与设备 [M]. 2版. 北京：冶金工业出版社，2001.

[3] 郑沛然. 炼钢学 [M]. 北京：冶金工业出版社，2004.

[4] 马竹梧，等. 钢铁工业自动化（炼钢卷）[M]. 北京：冶金工业出版社，2003.

[5] 周建男. 钢铁生产工艺装备新技术 [M]. 北京：冶金工业出版社，2004.

[6] 姜钧普，等. 钢铁生产短流程新技术——沙钢的实践（炼钢篇）[M]. 北京：冶金工业出版社，2001.

[7] 赵智. 电弧电极调节系统的预测控制 [D]. 沈阳：东北大学，2010.

[8] 贺庆，郭征. 电弧炉炼钢强化用氧技术的进展 [J]. 钢铁研究学报，2004（05）：1-4，50.

[9] 吴遵生，董渭峰，董文忠. 电弧炉偏心底出钢技术的实践与应用 [J]. 工业加热，2003（05）：60-63.

[10] 朱荣，张志诚，仇永全. 电弧炉炼钢炉壁碳氧喷吹系统的开发和应用 [J]. 特殊钢，2003（05）：39-40.

[11] 李杨文. 电弧炉用耐火材料在过去十年间的技术发展 [J]. 国外耐火材料，2003（05）：1-5.

[12] 丁秀中，王恭亮. 电炉偏心底出钢技术的应用 [J]. 冶金能源，2002（05）：23-25，38.

[13] 阎立懿. 现代超高功率电弧炉的技术特征 [J]. 特殊钢，2001（05）：1-4.

[14] 黄辉，邓彪，赵克猛. 30t 电弧炉偏心底出钢（EBT）改造及应用效果 [J]. 特钢技术，2000（01）：50-53.

[15] 田守信. 电弧炉出钢口用耐材的发展趋势 [N]. 中国冶金报，2012-04-05（B04）.

[16] 张艳利，王宪，贾全利. 炼钢电炉用耐火材料的现状和发展 [J]. 耐火与石灰，2019，44（03）：7-13.

[17] 薛娜，蓝振华，高留虎. 超高功率直流电炉用耐火材料维护小议 [C]. 中国金属学会耐火材料分会、中钢集团洛阳耐火材料研究院、先进耐火材料国家重点实验室、中钢科德孵化器（天津）有限公司、耐火材料杂志社. 2019年全国耐火原料学术交流会论文集，2019：331-336.

[18] 张艳利，王宪，贾全利. 炼钢电炉用耐火材料的现状和发展 [C]. 中钢集团洛阳耐火材料研究院有限公司、中国金属学会耐火材料分会、耐火材料杂志社、先进耐火材料国家重点实验室. 2018国际耐火材料学术会议论文集，2018：148-155，373-382.

[19] 王艳春，李魁猛，顼晓梅，等. 提高 40t 电弧炉间歇生产炉衬寿命的实践 [J]. 铸造技术，2018，39（05）：1033-1035.

[20] 李树民，陈兴润，杨丽敏，等. 电弧炉炉龄工艺研究与实践 [J]. 世界钢铁，2014，14（06）：10-14.

[21] 吴海，马杰. 电弧炉炼钢发展循环经济的研究 [J]. 甘肃冶金，2012，34（02）：25-29.

[22] 严永亮. 直流电弧炉炉底阳极用耐火材料 [J]. 宝钢技术，2008（01）：72-76.

[23] 王振宙，朱荣，蒋金燕，等. 电弧炉炉壁吹氧对渣线 MgO-C 砖的侵蚀分析 [J]. 冶金能源，2008（01）：45-46.

[24] 魏同，吴运广. 我国耐火材料生产节能方向 [J]. 耐火与石灰，2007（01）：4-8.

[25] 林涤凡. 提高电弧炉炉衬寿命的措施 [J]. 铸造，2006（11）：1198-1200.

[26] 曾玉清. 高效炼钢电弧炉设备 [J]. 工业加热，2004（01）：57-59.

[27] 田杭亮. 电弧炉节能技术 [J]. 工业加热，2002（06）：28-32.

[28] 蒋久信, 张国栋, 李纯, 等. 石墨对 MgO-C 耐火材料导电性能的影响 [J]. 耐火材料, 2002 (06): 329-332.

[29] 韩曙光. 60t DC 电弧炉导电炉底维护技术的改进 [J]. 南方金属, 2002 (05): 19-22.

[30] 桂明玺. 电炉用耐火材料的损毁 [J]. 国外耐火材料, 2002 (03): 40-44.

[31] 邢守渭. 炼钢电炉用耐火材料 [J]. 工业加热, 1997 (03): 1-5.